T0212768

SpringerBriefs in Computer Science

More information about this series at http://www.springer.com/series/10028

Deepak P · Prasad M. Deshpande

Operators for Similarity Search

Semantics, Techniques and Usage Scenarios

 Springer

Deepak P
IBM Research
Bangalore
India

Prasad M. Deshpande
IBM Research
Bangalore
India

ISSN 2191-5768   ISSN 2191-5776   (electronic)
SpringerBriefs in Computer Science
ISBN 978-3-319-21256-2   ISBN 978-3-319-21257-9   (eBook)
DOI 10.1007/978-3-319-21257-9

Library of Congress Control Number: 2015944152

Springer Cham Heidelberg New York Dordrecht London

Printed on acid-free paper

Springer International Publishing AG Switzerland is part of Springer Science+Business Media (www.springer.com)

Preface

With the growing variety of entities that have their presence on the web, retrieving relevant entities for various user requirements becomes an important problem. The area of Similarity Search that addresses this problem has received a lot of attention in the last fifteen years. Increasingly sophisticated data representations, query specifications, indexing mechanisms and algorithms to retrieve relevant entities to a query are being devised. Of these, developing indexes tailored to new kinds of data and devising algorithms to use such indexes to reduce the turnaround time for similarity search has attracted attention from the database community, resulting in several focused surveys and a few books that educate the audience about the field. Though relatively less discussed, another dimension in retrieval that has recorded tremendous progress over the years has been the development of mechanisms to enhance expressivity in specifying information needs. Similarity operators seek to advance the utility of similarity search systems from the user side by allowing the user to express her needs better by providing a richer set of querying options. In this book, we focus on the vocabulary of similarity operators that has grown vastly from just a set of two operators, top-k and skyline search, as it stood in the early 2000s. Today, there are ways to express complicated needs such as finding the top-k customers for a product wherein the customers are to be sorted based on the rank of the chosen product in their preference list. Some representative operators that have been proposed recently include K-N-Match, Reverse Furthest Neighbor, KN Diverse Neighbors, and Reverse kNN/Skyline operators. Arguably due to the complexity in the specification of new operators such as the above, uptake of such similarity operators has been low even though emergence of complex entities such as social media profiles warrants significant expansion in querying expressivity. To address this gap, we systematically survey the set of similarity operators, primarily focusing on their semantics, while also touching upon mechanisms to process them effectively. The aims of this book are to cover the following:

- A gentle introduction to the field of similarity operators starting from the fundamentals of similarity search systems.

- A comprehensive survey of the various similarity operators that have been proposed so far, in a structured manner to allow for easy assimilation.
- Positioning of the state-of-the-art in similarity operators with respect to the variety and complexity of entities that similarity search systems of today deal with, highlighting new directions and potential research gaps.
- A high-level overview of the indexing techniques and algorithms used for various types of data and similarity operators.

In this book, we expect to cover most of the important research advances in the area of similarity operators over the last fifteen years. To the best of our knowledge, this would be the first book focusing on the area of similarity operators.

The main emphasis of the book would be on providing a detailed tutorial on the area of operators for similarity search. The book will start off by providing introductory material on similarity search systems, highlighting the central role of similarity operators in such systems. This will include the insights gained from psychology and cognitive research that sheds light on the way the brain processes similarities and sets the stage for defining appropriate similarity measures and operators. This will be followed by a systematic categorized overview of the variety of similarity operators that have been proposed in literature over the last two decades. Indexing is a core technology to aid practical implementation of similarity operators; we will introduce and describe some of the indexing mechanisms that have been proposed in literature. Lastly, we will outline the research challenges in this area, so as to enable the interested researcher to identify potential directions of exploration. In summary, this book would provide a comprehensive overview of the field of similarity search operators, and cover the entire spectrum of technical issues related to the area.

We expect that this book would be useful for people across a wide variety of profiles such as students, educators and researchers. For students, we expect this would provide enough background to undertake research and implementation projects. In particular, for students who would like to build systems to illustrate the applicability of specific operator in the context of a specific domain or application, this book would provide a self-contained reference material. For educators who design and offer advanced graduate level courses on similarity search and recommender systems and would like to incorporate a segment on similarity operators, our book would be useful reference material to use as a platform for teaching and also to suggest as reading material for students. Researchers who are interested in advancing the state-of-the-art in similarity search would find this book useful to get up to speed to start working in the area. We have also included potential research directions for advancing this area keeping a research audience in mind.

We intend to start descriptions from the ground-up in a manner that would make the book accessible to undergraduate students with some familiarity with similarity search systems. Since similarity search systems and recommender systems are used by most people on a day to day basis, be it in the context of product search or social media browsing, we expect that most people with an interest in computer science would be able to understand this book. The chapter of indexing, however, expects the reader to have some background in information management; a person who has taken an undergraduate level course in databases should be able to easily grasp the contents of that chapter.

We hope you do enjoy and benefit from this book, and look forward to receiving any suggestions or comments that you might have at our respective email addresses. **Usage in Courses:** A detailed treatment of similarity operators covering all the content in this book would form a full *one-semester course*. Similarity Search, despite being a fairly active field in itself, is intimately related to various other disciplines such as recommender systems, information retrieval and case-based reasoning. Thus, selected parts of this book may be used as segments within courses focused on any of the above areas. We now outline some possible segments that could be carved out of this book:

- **Psychological Notion of Similarity:** Section 1.1 can be used as an introduction to the mind's notion of similarity, with a focus on those aspects that would potentially matter to similarity search systems.
- **Introduction to Similarity Search Operators:** Contents across Chapters 2 and 3 form an introductory segment to similarity operators.
- **Understanding Similarity Search Operators:** A segment covering Chapters 2, 3 and 4 can be covered to provide a reasonably detailed overview of similarity search operators to learners.
- **Deep-dive into Similarity Search Operators:** The meat of the book, comprising Chapters 2, 3, 4 and 5 form a comprehensive overview of the discipline of similarity search operators.
- **Indexing for Similarity Search:** Chapters 2, 3 and 6 contain material suitable for an introductory overview of similarity search with a focus on indexing.

Relevant sections of Chapter 7 may be added to each of the above segments for a graduate course to provide learners with a flavor for the kinds of research issues in similarity search systems that they may consider exploring in a course project or a larger research effort.

Online Resources: The website at *https://sites.google.com/site/ofssbook/* will host supporting content for this book such as powerpoint slides, and web links for additional reading.

Acknowledgements: The authors would like to acknowledge the support that they received from their friends and colleagues in the course of the planning and preparation of the material for this book. Deepak takes this opportunity to acknowledge the support and patience from his wife, Amrutha Jyothi, especially since most of the effort towards authoring this book was concentrated on the evenings and weekends. Prasad would like to express special thanks to his family, specifically his wife and kids for giving him the liberty to spend a signficant amout of family time on this project.

Bangalore,   *Deepak P*
May 2015   *Prasad M. Deshpande*

Contents

Chapter 1
Introduction

This (sic) sense of sameness is the very keel and backbone of our thinking... the mind makes continual use of the notion of sameness, and if deprived of it, would have a different structure from what it has. - William James, *Principles of Psychology*, 1890

In the last twenty years, the internet has become the medium to search for a large variety of objects that humans care about. This spectrum ranges from objects that are purely electronic (and intangible) such as social media profiles to objects such as products of various kinds whose digital representations are held in a database system. Object representations in relational database systems follow the schema-oriented approach, where a set of attributes is pre-defined for a database. It may be noted that we use the term object in this book to refer to a thing that potentially has an independent existence and is of interest to model; in particular, this notion is different from the abstraction used to refer to an encapsulation of data and methods as is commonly used in contexts such as object-oriented programming. A database of people records could have attributes such as *age*, *location* and *nationality*. Relational databases store data in tables where the attribute names form column labels, and each object is a row in the table; thus, an object usually takes one value for each attribute, which is stored in the corresponding cell. Traditional databases were mostly built to support *exact search* methods, where the query specifies a certain value for an attribute and processing of the query returns all objects in the database that have the same value for the attribute as that in the query. An example query would be to retrieve all people records where *nationality* is *"Indian"*. A simple extension would be to include ranges in case of ordinal attributes such as a query to retrieve all people records where the *age* is in the range *25 to 35*. More sophisticated search methods require joining multiple tables such as people and country tables to identify people who are located in a country that satisfy a specified threshold on the per capita income. The focus of this book, however, is on a different kind of querying over databases, called *similarity searching*.

Over time, to incorporate usage scenarios that cannot be addressed by exact search methods, methods for *proximity search* or *similarity search* have emerged. Similarity Search systems maintain, in addition to the data held in traditional database systems, a notion of similarity or distance between values for each at-

© The Author(s) 2015
D.P and P.M. Deshpande, *Operators for Similarity Search*,
SpringerBriefs in Computer Science, DOI 10.1007/978-3-319-21257-9_1

tribute. This may be in the form of a simple mathematical function as may be appropriate in the case of age, where the similarity between two age values may simply be quantified as a measure that is inversely related to the absolute numeric difference between them. In case of attributes such as nationality, a more complex domain-specific similarity measure may be necessary to incorporate intuitive notions such as the high similarity between *Indians* and *Sri Lankans* and the low similarity between *Indians* and the *British*. While it may be fairly easy to procure and model similarities between values of an attribute, users of a similarity search system are typically interested in similarity at the object level. Thus, a fundamental building block of similarity search systems is a means of aggregating attribute-based similarities to some measure at the object level. Typical operation of a similarity search system involves specifying a *query object*, and some additional parameters, upon which the similarity search system typically finds objects in the database that are *similar* to the query object. A simple, but computationally inefficient method would be to compare the query object with each object in the database to derive a quantification of similarity, and then, choosing a subset based on certain criteria to be output as results. The main source of variability among similarity search operations are in two phases:

- **Pairwise Similarity:** Similarities between the values of individual attributes may be obtained by using mathematical functions and/or domain-specific maps, as outlined earlier. A simple way to aggregate these attribute similarities to the object level would be to sum up the similarities across attributes to get an estimate of pairwise object similarity. However, in some cases, the user may want certain attribute similarities to be emphasized more than others; for example, in a health insurance application system, age may be a more important attribute. These could be incorporated by using a weight distribution between attributes, or allowing the user to specify the relative weighting between attributes at the query time. Instead of aggregating to a single value, one could choose to keep a summary representation of the attribute-wise similarity in order to avoid losing distributional information completely. Certain other scenarios require to keep the attribute-wise similarities separate without aggregation, as we will see later.
- **Result Set Computation:** Having estimated the pairwise similarity summary between the query and all objects in the database, the problem is now to choose the subset of objects to be output as results. The most usual case is to simply choose the top-k most similar objects to the query, where k is a user-specified parameter at query time. However, choosing the top-k is tricky in the case where the summary information is more than a single numeric value and other mechanisms to choose a subset of objects are needed. As in the case of the pairwise similarity computation phase, numerous possibilities exist in this phase as well.

In addition to the attribute set (aka schema) and the attribute-wise similarity estimation function, a similarity search system is fully specified once the choices for the above two phases are finalized. It may be noted that some similarity search methods such as multi-object queries and reverse queries do not strictly fall into this framework and require some extensions; nevertheless, this framework enables easy understanding of most common modalities of similarity search. The combination of the choices for summarizing pairwise object similarity and using it to estimate the result set is often referred to as the **similarity operator**. In this book, we will start with the basics of similarity operators, and study the various kinds of operators that have been surveyed in literature. The most commonly used and well-studied method for estimating pairwise similarity is the weighted sum approach, for which the result set if typically estimated as the top-k most similar objects; this is usually simply called the kNN or top-k operator. These were first studied in the context of multi-dimensional spatial data [12, 1, 9]. Newer operators started to appear in the database community since the 2000s; Table 1.1 shows an approximate timeline of when various operators were proposed in literature, listing a sample of *25* prominent operators.

We have tried to capture the most prominent similarity operators in Table 1.1. It may be noted that we are focusing on query-based retrieval methods; thus, the work on preference retrieval in databases (e.g., [4]), where the task is to retrieve objects based on how they fare in accordance with a scoring function that is fully specified only at run-time, have been excluded since their retrieval model is not query-specific. On the other hand, some operators that were proposed to rank objects using an implicit query (typically, assumed to be at the origin point in a Euclidean space, e.g., [19, 3, 23]) have been included since their semantics can be easily and meaningfully extended to use a different user-specified query point.

1.1 The Mind's Similarity Engine

The field of similarity search is about building systems that have an internal model of similarity that is closest to the model in the human mind[1] Similarity Operators may be seen as catering to the contextual nature of similarity in the mind; for example, a user who is interested in finding people with similar profiles to his own may be better off using the top-k operators, whereas an entrepreneur trying to find propsective customers for a planned restaurant may find the reverse top-k operator more useful. Thus, different similarity operators may be useful in different (explicit or implicit) contexts. The computing community has focused on building similarity search systems using the object model as introduced earlier, where objects to be compared have the same set of attributes, but take different values for each attribute. Thus, the high-level architecture of the traditional similarity search model is that of computing the attribute-wise similarity between the query and data objects, followed by using the chosen similarity operator to determine a result set. One of the efficiency-motivated assumptions that has been used extensively is that the attribute-wise similarity measure agrees to metric properties such as triangle inequality[2]. Another common assumption is that objects to be compared intuitively

---

[1] Apart from accuracy, another concern is efficiency, i.e., to fetch the results as quickly as possible. However, we will not worry about efficiency until a later chapter.

[2] http://www.scholarpedia.org/article/Similarity_measures - Accessed May 10th, 2015

Similarity Operator	Year	Pairwise Similarity	Result Set Estimation
Reverse Nearest/Farthest Neighbor [15]	2000	Any aggregated distance	Objects having query as nearest/farthest neighbor
Constrained NN [8]	2001	Any aggregated distance	Constraint on object position
Skyline [2]	2001	No Aggregation	Non-dominated Objects
Reverse kNN [29]	2004	Any aggregated distance	Objects having query among k nearest neighbors
K Nearest Diverse Neighbors [13]	2004	Any aggregated distance	K Nearest neighbors with diversity constraint on results
Thick Skyline [14]	2004	Any aggregated distance to compute ε neighbors	Non-dominated objects and their close neighbors
Constrained Skyline [23]	2005	No Aggregation	Skyline Operator restricted to objects satisfying a constraint
Dynamic Skyline [23]	2005	No Aggregation	Skyline Operator on a transformed space
Group-by Skyline [23]	2005	No Aggregation	Skyline Operator separately applied on groups specified on non-skyline attributes
Skyband [23]	2005	No Aggregation	Objects dominated by at most k objects
Top-k Frequent Skylines [3]	2006	No Aggregation	Top-k skyline points ranked acc to the number of sub-spaces in which they are part of skyline
K-N-Match [30]	2006	Similarity on N^{th} most similar attribute	K Nearest Neighbors based on N^{th} most similar attribute
Top-k Representative Skyline [19]	2007	No Aggregation	Top-k Skyline points such that number of objects collectively dominated is maximized
Reverse Skyline [5]	2007	No Aggregation	Objects where query is not dominated w.r.t them
Visible Nearest Neighbors [22]	2007	Any aggregated distance	Nearest Neighbors that are visible to the query
Subspace Top-k/Range [18]	2008	Distance aggregated over chosen attribute subset	Top-k closest objects to query, or those that satisfy range constraint
Reverse-k Farthest Neighbor [16]	2008	Any aggregated distance	Objects having query among k farthest neighbors
Visible RkNN [10]	2009	Any aggregated Distance	Objects having query among k visible nearest neighbors
Multi-Query Top-k [28]	2012	Weighted Sum aggregate distance across attributes and queries	Top-k closest objects on aggregated distance
Skyline k-Groups [17]	2012	Any aggregation to compute group vector	Groups of k objects not dominated by other groups
Reverse k-Skyband [20]	2012	No Aggregation	Objects having query among their k-Skyband
Range Reverse NN [24]	2013	Any aggregated distance	Objects with any point in the query area as nearest neighbor
Hypermatching Top-k [33]	2014	Aggregated distance with amplification for extreme values	k objects most similar to query
Reverse k-Ranks [34]	2014	Any aggregated distance	Objects ranked based on the rank of the query with respect to them
Geo-social Skyline [6]	2014	No aggregation	Objects in the skyline that are graph-proximal to query

Table 1.1: Similarity Operators Timeline

have the same schema (i.e., the set of attributes). Yet another assumption is that the pairwise similarity between two objects is not influenced by a third object. Given the complexity of the cognitive process in the human mind, it has been shown that many of these assumptions do not hold many a time. Having said that, the large success of similarity search systems probably could be used to say that the traditional model is a useful proxy for the complex similarity engine of the human mind. On the other hand, the proliferation of similarity operators in recent years points to the fact that there are scenarios that are not adequately addressed by simple operators. A better understanding of the observations on the mind's similarity engine, we think, would be useful to those interested in advancing the frontier in similarity search systems, and similarity search operators in particular. Towards this, we present a brief glimpse of the literature from psychology on the complexity of similarity estimation within the human mind.

1.1.1 Metric Properties

Most literature in similarity search systems assume that the attribute-specific pairwise distance/dissimilarity measure, $d(.,.)$ is a metric, which implies that it satisfies a set of four properties. To best illustrate this, consider various values of an attribute as $\{v_1, v_2, \ldots\}$. For an attribute such as *City*, these could be values such as *Bangalore*, *Chicago* and *Sydney*. The metric condition requires that the following hold:

- **Self-Dissimilarity:** This says that the self-dissimilarity of any value is equivalent to that of any other. Formally, $\forall x, y, d(v_x, v_x) = d(v_y, v_y)$. Typical distance functions such as that for distance between cities may assign the self-dissimilarity to 0.0.
- **Minimality:** The minimality condition requires that the dissimilarity between any two distinct objects be higher than the self-dissimilarity. This may be expressed as $\forall x, y, z \; x \neq y, d(v_x, v_y) \geq d(v_z, v_z)$. For our example of cities, this is intuitive since the distance between two different cities would be greater than 0.0.
- **Symmetry:** This condition enforces that the dissimilarity function be symmetric and independent of the order of objects. Thus, $\forall x, y, d(v_x, v_y) = d(v_y, v_x)$. Though intuitive, this condition may not necessarily hold if there are one-way streets between cities.
- **Triangle Inequality:** Triangle inequality restricts the dissimilarity between any two values to be at most the sum of their mutual distances from a third. In formal notation, $\forall x, y, z, d(v_x, v_y) \leq d(v_x, v_z) + d(v_y, v_z)$. For example, if v_x happens to be very proximal to v_z and v_z is very proximal to v_y, then, v_x needs to be proximal to v_y too. Continuing with our cities example, the distance between any two cities may be estimated as the minimum of the lengths of all paths that connect the two cities; since the path through a third city also forms a legitimate path, this can evdiently be at best the shortest path between the cities.

Similarity functions are usually inversely related to distance functions; a simple and commonly used transformation [27] is the following:

$$s(v_x, v_y) = e^{-d(v_x, v_y)} \tag{1.1}$$

Thus, constraints on the $d(.,.)$ function would transform to analogous constraints on the $s(.,.)$ function. Though most of the metric conditions seem intuitive, it so happens that they do not necessarily hold for the mind's similarity engine. We will now describe some studies that help illustrate cases where the metric properties do not necessarily hold.

Self-Dissimilarity

In the late 1970s, Podgorny and Garner [25], both with the Yale University, conducted a study to instrument the human perception of visual similarity between different english letters. Their simple experiment measured reaction time of different subjects for a binary-choice task of estimating whether two symbols shown to them are of the same letter in the alphabet. Reaction time had been shown to be a measure correlated with confusability, and thus, a good proxy of dissimilarity between the two symbols. To measure self-dissimilarity, the same letter was shown twice (as two elements of a pair) and the reaction time for determining the *yes* answer was measured. The result of this experiment supported the surprising conclusion that every object is not equally dissimilar to itself. For example, the mean reaction times for the pair $[D, D]$ was found to be $490ms$ whereas the reaction times for the $[H, H]$ pair was recorded to be $563ms$. There were many other alphabet pairs where the reaction times varied widely. Thus, there are simple cases such as visual character recognition where the self-dissimilarity axiom does not hold.

Minimality

In the same study by Podgorny and Garner referenced above, another experiment was conducted to check whether a symbol is indeed more similar to itself, than other different symbols among themselves. Towards this, two alphabet sets were formed $\{C, D, O, Q\}$ and $\{H, N, M, W\}$; these were empirically identified as sets with very high intra-set similarity. The volunteers for the experiment were told about these sets, and subjected to the same binary choice experiment, this time at the level of sets. Thus, symbols were required to be judged based on whether they belong to the same set or not. For example, the $[C, D]$ pair is expected to elicit a *yes* response whereas the correct response for the $[D, W]$ pair would be a *no*. Since response times are regarded to be a proxy of dissimilarity, the minimality condition translates to the following:

$$\forall v_x, v_y, v_z, \text{ such that } v_x \neq v_y, Set(v_x) = Set(v_y), RT(v_x, v_y) > RT(v_z, v_z) \tag{1.2}$$

Thus, a symbol would need to have a smaller reaction time with itself, than a pair of different symbols who belong to the same set. As in the case of the self-dissimilarity axiom, the experiment confirmed that minimality also need not necessarily hold. The pair $[C,D]$ elicited a *yes* response in $527ms$ whereas the *yes* response for the $[H,H]$ pair took an average of $563ms$. Thus, C is regarded to be more similar to D than H is to itself. Similarly, there were other cases for which the minimality condition did not hold. Further evidence of the violation of the minimality condition emerged from another late 1970s study conducted at Johns Hopkins University [11] on a similar task of character recognition where subjects were shown various english letters for a very brief time (10-70 ms) and asked to identify them. The stimulus (i.e., visual input) was adjusted in duration so that subjects' performance agree to a pre-decided error rate. Surprisingly, the symbol M was more often recognized as the letter H (probability of 0.391) and N (0.165) than as M itself (0.11). These two studies confirm that the minimality condition need not necessarily hold in general cases.

Symmetry

The symmetry condition requires that similarity between objects be assesed regardless of the order in which they are presented. A pioneering study by Amos Tversky [31] illustrated the violation of symmetry in similarity judgements between common entities such as countries. In particular, when asked to pick one sentence among *North Korea is similar to Red China* and *Red China is similar to North Korea,* 66 out of 69 participants preferred the former. This asymmetry was further evident in a study where people were asked to judge similarities between pairs of countries in two groups, where only the order of the pair differed across groups; the difference between their judgements were found to be statsitically significant at a very convincing p-value of < 0.01. Based on the inferences from his study, Tversky observes that people are generally inclined to assess similarity highly when a less prominent object is presented first with a similar and more prominent object.

Triangle Inequality

Tversky and Gati [32], in 1982, presented an early study that illustrates that the combination of triangle inequality and segmental additivity[3] are violated many a time. Infact, many of the common similarity measures that are used in similarity search systems of today do not satisfy the triangle inequality. Some examples of distance measures that violate the triangle inequality are given below:

---

[3] Segmental additivity postulates that if A, B and C lie on a staight line with B being in between A and C, $d(A,B) + d(B,C) = d(A,C)$ holds.

Fig. 1.1: Multi-modal Similarity Example (from [26])

- Dynamic Time Warping: This is a very popular distance measure for time series matching. An example illustrating the violation of triangle inequality in DTW appears in [21].
- KL-Divergence: This popular measure to compare probability distributions, apart from not satisfying triangle inequality, isn't even symmetric and depends on the order of the parameters in the input.
- Cosine Similarity: Cosine similarity is the similarity measure of choice for comparing text documents in domains such as information retrieval, that quantifies the similarity between two vectors by the cosine of the angle between them. The simple distance function that takes the difference between the cosine similarity and unity does not satisfy the triangle inequality either.
- Pattern Matching: Dissimilarity measures between patterns such as non-linear elastic matching [7] do not adhere to the triangle inequality property.

In conclusion, while developing new similarity operators and similarity functions for scenarios so far inadequately addressed in literature, one need not feel constrained by the need to respect the metric properties since similarity measures of the mind do not necessarily agree to such properties.

1.1.2 Multi-modal Similarity Search

Objects made up of text data such as text documents and/or text snippets are often represented using their lexical content. On the other hand, images are often represented using the objects they contain, borders of the objects, or using lower-level

 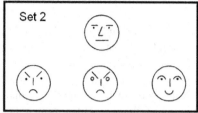

Fig. 1.2: Similarity - Grouping Inter-relatedness (from [31])

representations such as distribution of pixels across different colors. Thus, traditional representations of text and images seldom have any attribute in common and come from very different schemas, making the traditional attribute-based comparison model inapplicable. However, the mind still has a notion of multi-modal similarity across text and images that is not necessarily dependent on the correlation between the meaning of the text and the object in the image. An illustrative example from [26] appears in Figure 1.1 that illustrates four objects: a star shaped image with sharp edges, an ameobic image with rounded edges, and two words *bouba* and *kiki*. When human subjects were asked to choose an image to text mapping, the choice was found to be unanimous in the sense most people chose to associate *kiki* with the star-shaped figure and *bouba* with the ameobic figure. The visual similarity between the acute angles in *k* with the star-pattern and the correlation between the rounded edges of *b* and *o* with those of the right figure were apparently not the reason behind such a mapping, since the mapping was consistent even when experimented with other natural languages where visual similarities were non-existent. The author of [26], VS Ramachandran, proposes the following hypothesis as the reason behind the mapping:

- The smooth rounding and relaxing of the lips when pronouncing *bouba* is correlated well with the rounded edges in the right figure.
- The sharp inflection of the tongue on the palate when uttering *kiki* has similarities with the sharp edges in the star-shaped figure.

While it could be possible that one could come up with other hypotheses to explain the mapping, it seems apparent that any matching function using most of the traditional representations of text and images would be unable to identify such a correlation. Thus, similarity search across different modalities pose a problem in defining the schema as well.

1.1.3 Similarity and Grouping: The Diagnosticity Principle

We will now revisit the work by Tversky [31] to illustrate another anomaly of interest in similarity searching. It is fairly intuitive to assume that the similarity between a pair of objects depends on only them, and not on others; this is a very common assumption used in virtually every similarity search system of today. Tversky reports a study that shows very compelling violations of the principle in a study using a visual depiction of objects as illustrated in Figure 1.2. First, consider Set 1 where the top smiley is to be matched with the three smileys below it; human subjects are posed with the task of identifying the most similar smiley to the one on the top. The study recorded the distribution of votes to be [44%, 14%, 42%]; thus, the left smiley is the most preferred one. The second task was of the same nature, now posed on Set 2. On this task, the distribution of votes from left to right was found to be [12%, 8%, 80%]. Now, it is useful to pay attention to the differences between the sets; it should be fairly easy to recognize that the only difference is that of the change in the middle smiley from the smiling face in Set 1 to the frowning face in Set 2. Thus, the switch between the preference to the left smiley in Set 1 to the right smiley in Set 2 is caused by the change in the middle smiley, the one that is not the preferred one in either of the sets. The explanation that Tversky proposes is that there is a tendency to group objects (here, smileys) in the mind; thus, in Set 1, the smiling faces get grouped together, and the one left out is preferred to be judged more similar to the query. In Set 2, the frowning faces get grouped, thus the smiley that gets left out happens to the smiling face, which gets a higher fraction of votes consequently. In effect, the judgements of the similarity between the query and the left and right smileys are heavily influenced by the middle smiley; this inter-relatedness between similarity assesment and grouping is referred to as the diagnosticity principle. Though well accepted in psychology literature, this notion has largely been unexplored in the field of computing despite being pertinent in areas such as similarity search. It is worth noting, in this context, that a recent work [33] tries to incorporate such notions in similarity estimation by exaggerating the importance of extreme values for an attribute as we will see in a later chapter.

1.2 Summary

In this chapter, we considered the operation of common similarity search systems, more from a semantics point of view as opposed to the efficiency-oriented view as used in typical database literature. We illustrated that the full-specification of a similarity search system involves the schema definition as well as the phases of pairwise similarity estimation and result set identification. Variations in the specification of pairwise similarity estimation and result set identification give rise to various similarity operators. In addition to reviewing the most common similarity operator, the top-k operator, we looked at the landscape of similarity operators that have been proposed in the last two decades. Despite the proliferation of similarity operators, the discipline of similarity search still remains a fertile area for research. We looked at some disconnects between assumptions made in similarity search systems and conventional wisdom from psychology wherein we discussed simple scenarios that are known to violate many of the assumptions made in the computing literature. We hope this sets the context for further discussion on similarity search operators in the subsequent chapters.

References

1. J. L. Bentley. Multidimensional binary search trees used for associative searching. *Commun. ACM*, 18(9):509–517, 1975.
2. S. Borzsony, D. Kossmann, and K. Stocker. The skyline operator. In *Data Engineering, 2001. Proceedings. 17th International Conference on*, pages 421–430. IEEE, 2001.
3. C.-Y. Chan, H. Jagadish, K.-L. Tan, A. K. Tung, and Z. Zhang. On high dimensional skylines. In *Advances in Database Technology-EDBT 2006*, pages 478–495. Springer, 2006.
4. Y.-C. Chang, L. Bergman, V. Castelli, C.-S. Li, M.-L. Lo, and J. R. Smith. The onion technique: indexing for linear optimization queries. In *ACM SIGMOD Record*, volume 29, pages 391–402. ACM, 2000.
5. E. Dellis and B. Seeger. Efficient computation of reverse skyline queries. In *Proceedings of the 33rd international conference on Very large data bases*, pages 291–302. VLDB Endowment, 2007.
6. T. Emrich, M. Franzke, N. Mamoulis, M. Renz, and A. Züfle. Geo-social skyline queries. In *Database Systems for Advanced Applications*, pages 77–91. Springer, 2014.
7. R. Fagin and L. Stockmeyer. Relaxing the triangle inequality in pattern matching. *International Journal of Computer Vision*, 30(3):219–231, 1998.
8. H. Ferhatosmanoglu, I. Stanoi, D. Agrawal, and A. El Abbadi. Constrained nearest neighbor queries. In *Advances in Spatial and Temporal Databases*, pages 257–276. Springer, 2001.
9. R. A. Finkel and J. L. Bentley. Quad trees: A data structure for retrieval on composite keys. *Acta Inf.*, 4:1–9, 1974.
10. Y. Gao, B. Zheng, G. Chen, W.-C. Lee, K. C. Lee, and Q. Li. Visible reverse k-nearest neighbor queries. In *Data Engineering, 2009. ICDE'09. IEEE 25th International Conference on*, pages 1203–1206. IEEE, 2009.
11. G. Gilmore, H. Hersh, A. Caramazza, and J. Griffin. Multidimensional letter similarity derived from recognition errors. *Perception & Psychophysics*, 25(5):425–431, 1979.
12. A. Guttman. R-trees: A dynamic index structure for spatial searching. In *SIGMOD'84, Proceedings of Annual Meeting, Boston, Massachusetts, June 18-21, 1984*, pages 47–57, 1984.
13. A. Jain, P. Sarda, and J. R. Haritsa. Providing diversity in k-nearest neighbor query results. In *Advances in Knowledge Discovery and Data Mining*, pages 404–413. Springer, 2004.
14. W. Jin, J. Han, and M. Ester. Mining thick skylines over large databases. In *Knowledge Discovery in Databases: PKDD 2004*, pages 255–266. Springer, 2004.
15. F. Korn and S. Muthukrishnan. Influence sets based on reverse nearest neighbor queries. In *ACM SIGMOD Record*, volume 29, pages 201–212. ACM, 2000.
16. Y. Kumar, R. Janardan, and P. Gupta. Efficient algorithms for reverse proximity query problems. In *Proceedings of the 16th ACM SIGSPATIAL international conference on Advances in geographic information systems*, page 39. ACM, 2008.
17. C. Li, N. Zhang, N. Hassan, S. Rajasekaran, and G. Das. On skyline groups. In *Proceedings of the 21st ACM international conference on Information and knowledge management*, pages 2119–2123. ACM, 2012.
18. X. Lian and L. Chen. Similarity search in arbitrary subspaces under l p-norm. In *Data Engineering, 2008. ICDE 2008. IEEE 24th International Conference on*, pages 317–326. IEEE, 2008.

19. X. Lin, Y. Yuan, Q. Zhang, and Y. Zhang. Selecting stars: The k most representative skyline operator. In *Data Engineering, 2007. ICDE 2007. IEEE 23rd International Conference on*, pages 86–95. IEEE, 2007.

20. Q. Liu, Y. Gao, G. Chen, Q. Li, and T. Jiang. On efficient reverse k-skyband query processing. In *Database Systems for Advanced Applications*, pages 544–559. Springer, 2012.

21. M. Müller. Dynamic time warping. *Information retrieval for music and motion*, pages 69–84, 2007.

22. S. Nutanong, E. Tanin, and R. Zhang. Visible nearest neighbor queries. In *Advances in Databases: Concepts, Systems and Applications*, pages 876–883. Springer, 2007.

23. D. Papadias, Y. Tao, G. Fu, and B. Seeger. Progressive skyline computation in database systems. *ACM Transactions on Database Systems (TODS)*, 30(1):41–82, 2005.

24. R. Pereira, A. Agshikar, G. Agarwal, and P. Keni. Range reverse nearest neighbor queries. In *KICSS*, 2013.

25. P. Podgorny and W. Garner. Reaction time as a measure of inter-and intraobject visual similarity: Letters of the alphabet. *Perception & Psychophysics*, 26(1):37–52, 1979.

26. V. S. Ramachandran. *The tell-tale brain: A neuroscientist's quest for what makes us human.* WW Norton & Company, 2012.

27. R. N. Shepard. Toward a universal law of generalization for psychological science. *Science*, 237(4820):1317–1323, 1987.

28. Y. Shi and B. Graham. A similarity search approach to solving the multi-query problems. In *Computer and Information Science (ICIS), 2012 IEEE/ACIS 11th International Conference on*, pages 237–242. IEEE, 2012.

29. Y. Tao, D. Papadias, and X. Lian. Reverse knn search in arbitrary dimensionality. In *Proceedings of the Thirtieth international conference on Very large data bases-Volume 30*, pages 744–755. VLDB Endowment, 2004.

30. A. K. Tung, R. Zhang, N. Koudas, and B. C. Ooi. Similarity search: a matching based approach. In *Proceedings of the 32nd international conference on Very large data bases*, pages 631–642. VLDB Endowment, 2006.

31. A. Tversky. Features of similarity. *Psychological Reviews*, 84(4):327–352, 1977.

32. A. Tversky and I. Gati. Similarity, separability, and the triangle inequality. *Psychological review*, 89(2):123, 1982.

33. R. Yager and F. Petry. Hypermatching: Similarity matching with extreme values. *Fuzzy Systems, IEEE Transactions on*, 22(4):949–957, Aug 2014.

34. Z. Zhang, C. Jin, and Q. Kang. Reverse k-ranks query. *Proceedings of the VLDB Endowment*, 7(10), 2014.

Chapter 2
Fundamentals of Similarity Search

We will now look at the fundamentals of similarity search systems, providing the background for a detailed discussion on similarity search operators in the subsequent chapters. We first look at object representations used in similarity search systems, and then consider attribute types as well as similarity measures used for various attribute types.

2.1 Object Representation

As outlined in Chapter 1, similarity search systems most commonly use an attribute-based representation for the dataset of objects to perform similarity search on. Accordingly, a schema would consist of a listing of chosen attributes such as *age*, *name* and *address* along with the domain of values for each of the attributes. A dataset that is based on this schema would have many objects, each object taking a value for each of the specified attributes in the schema; *null* values may or may not be allowed depending on whether the similarity engine can estimate the similarity between a value and a *null*. Thus, a full specification of the data is achieved when the value for each attribute is specified for each object in the dataset. Despite the implied formalism in the data definition as discussed above, the pervasiveness of this representation on the web may be immediately obvious to the reader. Nevertheless, Figure 2.1 presents a collage of attribute based representations for different kinds of objects in the web such as gadgets, sports personalities and nations.

2.2 Attribute Types

The set of attributes that make up the schema could vary in the nature of values that they take. For example, the *age* attribute in a people database would typically take numeric values, whereas the *name* attribute would take short strings as values.

© The Author(s) 2015
D.P and P.M. Deshpande, *Operators for Similarity Search*,
SpringerBriefs in Computer Science, DOI 10.1007/978-3-319-21257-9_2

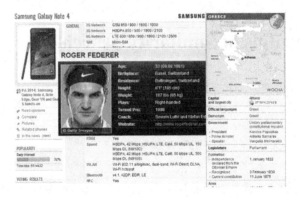

Fig. 2.1: Collage of Attribute-based Representation Screenshots

We will now take a brief look at the types of attributes that are commonly used in databases upon which similarity search systems work.

- **Numeric:** Readings from sensors such as those that measure climatic attributes such as temperature and pressure would naturally take on numeric values. The popular Iris dataset[1] of flowers models flowers in terms of their attributes such as *sepal* and *petal* lengths and widths, all of which are numeric attributes. The population attribute for a countries dataset would also take numeric values. While all of the above take single numeric values, there could be attributes that take multi-numeric values such as locations in a x-y co-ordinate space, where a pair of numeric attributes for a 2-tuple represent each location. Though not strictly numeric, the latitude-longitude representation of geographic locations are also similar in the sense that both come from a well-ordered set.
- **Categorical:** Categorical attributes are another popular type, which can take on one of a limited number of possible values. The *gender* attribute is a common example of a categorical attribute. Other attributes such as *education* that can take on values such as *undergraduate*, *graduate* etc. are also categorical attributes. For a database of computing servers, the *operating system* would be a categorical attribute.
- **Text:** People profiles could contain attributes such as *short bio* or *address* fields that are typically textual where similarity assesment could be done between values using text similarity measures.
- **Sequence/Time Series:** Genes and proteins are typically attributes that are represented using a sequence of bases or amino acids respectively. Other common examples of sequences include health information such as ECG or EEG. Data from any kind of sensors are also naturally time series data.

---

[1] http://en.wikipedia.org/wiki/Iris_flower_data_set - Accessed May [th], 2015

The above list of kinds of attribute data are merely meant to be illustrative of the variety in attribute types. As is probably obvious, this list is not meant to be comprehensive.

2.3 Comparing Attribute Values: Similarity and Distance

Computing the similarity between two values of an attribute is a fundamental operation in similarity search systems. As seen in the introduction chapter, attribute-wise similarities are then aggregated in a manner as determined by the similarity operator, to arrive at a pair-wise similarity measure for objects. Though glossed over earlier, it is useful to point out the inter-convertibility between the notions of similarity and distance in typical scenarios. As is evident from the name, similarity and distance (i.e., dissimilarity) are inversely related; two values that have a large distance between them are likely to be judged to score low on similarity and vice versa. Similarity measures are generally expected to be in the range $[0, 1]$ whereas many popular distance measures estimate distances to be any non-negative value. In the special case where distance is computed to be in the range $[0, 1]$, the conversion could be simply done by subtracting the distance from unity[2]:

$$s = 1 - d \qquad (2.1)$$

where s and d refer to similarity and distance respectively. For strictly positive distances, the reciprocal would serve as a measure of similarity. i.e.,

$$s = \frac{1}{d} \qquad (2.2)$$

For a general non-negative distance measure, division by zero could be avoided by adding unity to the denominator as follows:

$$s = \frac{1}{d + 1} \qquad (2.3)$$

Another possible transformation [4] that was alluded to in Chapter 1 uses exponentiation:

$$s = e^{-d} \qquad (2.4)$$

One may want to *control* this transformation in such a way that similarity be estimated to be 0.5 (50% of maximum similarity) at a particular value of distance. This could be done by using another parameter τ in the transformation such as below:

$$s = e^{\frac{-d}{\tau}} \qquad (2.5)$$

---

[2] http://people.revoledu.com/kardi/tutorial/Similarity/WhatIsSimilarity.html - Accessed January 20, 2015

Under this transformation, s would get estimated to 0.5 when the distance is $0.693 \times \tau$. Instead of choosing a pivot for 50% of the maximum similarity, it is straightforward to extend this to other pivots. A combination of transformations could also be used to convert distance to similarity; for example, one may use the sigmoid function[3] to transform the general non-negative distance measure to restrict it to the $[0, 1]$ range, and then use the subtraction transformation as in Equation 2.1.

In short, there are a variety of possible mechanisms to convert similarity to distance and vice versa. In this book, when we consider comparing values, we will describe distance or similarity based on what is more convenient to describe; for example, distance is a very natural measure to compare numbers, whereas vectors may be compared by measuring their cosine similarity. It would be left to the reader to do the transformation between distance and similarity as needed, such as for implementation purposes.

2.4 Similarity Measures

A similarity measure, or a similarity function, is an attribute-specific function that quantifies the similarity between two values of the attribute. In this section, we will briefly look at a few common similarity measures, for various kinds of attributes.

- **Numeric Attributes:** For simple numeric attributes such as *age*, the difference between two values could be used as a measure of distance/dissimilarity. For multi-numeric attributes such as a pair of (x, y) co-ordinates or a vector of values, the Minkowski distance metric is often used as a measure of distance. The Minkowski distance[4] between two vectors v_1 and v_2 would be estimated as:

$$\left(\sum_i (|v_1[i] - v_2[i]|)^p \right)^{1/p} \tag{2.6}$$

For $p = 1$, this is equivalent to the sum of the differences between the corresponding values (i.e., the Manhattan distance[5], whereas $p = 2$ yields the Euclidean distance [1]). When $p \to \infty$, the Minkowski distance is simply equal the to maximum of the element-wise distances.
- **String Attributes:** Two values of a string attribute that are equal in length could be compared using the Hamming distance [2], that simply counts the number of positions in which the strings have different characters. For strings that need not necessarily be equal in length, Levenshtein distance [3] is often used. The Levenshtein metric considers a set of three operations, deletion, insertion and substitution, and counts the minimum number of operations that are needed to transform one string to the other. The Levenshtein distance between two strings,

---

[3] http://en.wikipedia.org/wiki/Sigmoid_function - Accessed 21st January, 2015

[4] http://en.wikipedia.org/wiki/Minkowski_distance - Accessed 22nd January, 2015

[5] http://mathworld.wolfram.com/ManhattanDistance.html - Accessed 22nd January, 2015

a and b is estimated as $lev_{a,b}(|a|,|b|)$ that uses a recursive function as follows:

$$lev_{a,b}(i,j) = \begin{cases} max(i,j) & \text{if } min(i,j) = 0 \\ min \begin{cases} lev_{a,b}(i-1,j)+1 \\ lev_{a,b}(i,j-1)+1 \\ lev_{a,b}(i-1,j-1)+I(a[i] \neq b[j]) \end{cases} & \text{otherwise} \end{cases}$$

(2.7)

where $I(.)$ is the indicator function that evaluates to 1 and 0 when the inner condition is true and false respectively. The Damerau-Levenshtein distance[6] extends the Levenshtein distance function by allowing for an additional operation, that of transposition of adjacent characters. Consider two strings *rick* and *irck*; these have Hamming and Levenshtein distance both evaluating to 2.0 due to mismatches on the first two characters, whereas these could be made identical by just one transposition leading to them being judged as just a unit distance away from each other according to the Damerau-Levenshtein distance.

- **Text Attributes:** Text attributes are typically represented as bags of words, wherein the ordering of the words in the text data is discarded. For example, the text fragment *this is a dog* would be represented as $\{this = 1.0, is = 1.0, a = 1.0, dog = 1.0\}$; the value associated with each word denotes its frequency in the text fragment. Each text fragment could be thought of as a vector in a (highly) multi-dimensional space where each dimension corresponds to a word in the dictionary, with the value on each dimension being equal to the frequency of the corresponding word in the text fragment. Thus, our sample text fragment takes on non-zero values only on the four dimensions $\{this, is, a, dog\}$. Instead of using simply the word frequency, the popular tf-idf weighting scheme[7] uses the product of word frequency and the IDF, a measure inversely related to the popularity of the word in a collection of documents such as news articles, to associate each dimension with. Whatever be the vector representation, word frequency based or tf-idf based, two text fragments are generally compared using the cosine similarity measure[8] that quantifies the cosine of the angle between their vector representations in the multi-dimensional space. Two identical vectors would have 0° between them, and would thus be evaluated to have a similarity of unity.

- **Geographic Locations:** Geo-locations such as latitude-longitude pairs form another kind of popular attribute type. The distance between two points on the earth's surface is often estimated using the great circle distance, the shortest distance between two points along the great circle[9] that contains both the points. The Haversine formula[10] is a popular method to calculate the great circle distance.

- **Time-Series Data:** ECG, EEG and other kinds of temporally sequenced information form an integral part of personal health profiles; thus, similarity search for

---

[6] http://en.wikipedia.org/wiki/Damerau-Levenshtein_distance - Accessed January 22nd, 2015

[7] http://en.wikipedia.org/wiki/tf-idf - Accessed January 28th, 2015

[8] http://en.wikipedia.org/wiki/Cosine_similarity - Accessed January 28th, 2015

[9] http://en.wikipedia.org/wiki/Great_circle - Accessed January 28th, 2015

[10] http://mathforum.org/library/drmath/view/51879.html - Accessed January 28th, 2015

diagnostic services would need to estimate similarity between time-series data. Time-series data also are common in climate-related use cases where environmental factors such as temperature and humidity are measured over elongated periods of time. A simple method of comparing two time-series data would be to decide on a time interval - say, 5 seconds - and compare the values of the time series pair after the lapse of each interval. Thus, the values of the first time series at 5 seconds and 10 seconds from the start would be compared with the values of the second time series at 5 seconds and 10 seconds respectively. These pairwise comparisons would then be aggregated to arrive at a single similarity value for the time series pair. Though simple and appealing, this fixed interval comparison approach has several drawbacks, of which we will descibe one using an example. More often than not, there are periodicities in time-series data (e.g., ECG), and there are bound to be slight differences in the time intervals of these periodic patterns across time series; the fixed interval comparison method, due to usage of fixed intervals, would get out of sync and compare the crest of one time series against the trough of the other every now and then, resulting in exaggerated distance estimates. Dynamic Time Warping [5], among the more popular methods to compare time-series data, is robust to speed variations and uses a simple recursive formulation to compute a more realistic distance estimate.

2.5 Summary

In this chapter, we introduced the attribute-based object representation that is widely used for representing real-world entities on the web as well as in similarity search systems. This was followed by a brief description of various attribute types as well as measures used to quantify similarity and dissimilarity for specific attribute types. It was also noted that similarity and dissimilarity (distance) are often interchangeable through simple transformations, and that once either of them is specified, it is easy to derive the other.

References

1. M. M. Deza and E. Deza. *Encyclopedia of distances*. Springer, 2009.
2. R. W. Hamming. Error detecting and error correcting codes. *Bell System technical journal*, 29(2):147–160, 1950.
3. V. I. Levenshtein. Binary codes capable of correcting deletions, insertions and reversals. In *Soviet physics doklady*, volume 10, page 707, 1966.
4. R. N. Shepard. Toward a universal law of generalization for psychological science. *Science*, 237(4820):1317–1323, 1987.
5. X. Wang, A. Mueen, H. Ding, G. Trajcevski, P. Scheuermann, and E. Keogh. Experimental comparison of representation methods and distance measures for time series data. *Data Mining and Knowledge Discovery*, 26(2):275–309, 2013.

Chapter 3
Common Similarity Search Operators

In the previous chapter, we saw how objects are represented using attributes and how attribute wise distances are computed for different attribute types. These attribute-wise distances can be aggregated to compute distances between objects which can be used to identify objects satisfying a similarity query. There can be different notions of similarity depending on the context of the user and the objective of the search. In this chapter, we will first describe a framework that enables expression of different similarity operators as combinations of aggregation and filter functions. Further, we will look at some commonly used instances of these aggregation and filter functions.

3.1 The Similarity Search Framework

Consider a database of n objects denoted as $\mathcal{X} = \{x_1, x_2, \ldots, x_n\}$ defined on a schema of m attributes denoted as $\mathcal{A} = \{a_1, a_2, \ldots, a_m\}$. A full description of the data is achieved by specifying the value of each attribute of each data object, i.e., $\forall i, j, 1 \leq i \leq n, 1 \leq j \leq m, x_i.a_j$ where $x_i.a_j$ denotes the value of the jth attribute of the ith object. We will sometimes take the freedom to denote it by x_{ij} to avoid notation clutter. The domain of the jth attribute, Dom_j specifies the allowable set of values for the jth attribute of any data object that confirms to the schema $\mathcal{A}$. Apart from the dataset, similarity search systems also use a set of attribute-specific similarity functions. $s_j(.,.)$, a function from $Dom_j \times Dom_j \rightarrow \mathcal{R}$ denotes the similarity between two values of the jth attribute. The analogous dissimilarity function for the jth attribute would be denoted as $d_j(.,.)$. The data, schema, and (dis)similarity functions form the basic components of a similarity search system. Additionally, similarity search systems may pre-process data to create indexes to help speed-up processing of similarity search queries; but, we will not consider indexes until a later chapter. The notations used by us are summarized in Table 3.1.

The operation of the similarity search system runtime can begin when a specification of a query q and the choice of a similarity search operator $\mathcal{O}$ is made available.

© The Author(s) 2015
D.P and P.M. Deshpande, *Operators for Similarity Search*,
SpringerBriefs in Computer Science, DOI 10.1007/978-3-319-21257-9_3

Notation	Interpretation
$\mathcal{X}$	Dataset for Similarity Search
x_i	i^{th} object in $\mathcal{X}$
x	Any data object in $\mathcal{X}$
$\mathcal{A}$	Set of Attributes for objects in $\mathcal{X}$
m	Number of Attributes in $\mathcal{A}$
a_j	j^{th} attribute in $\mathcal{A}$
$x_i.a_j$	j^{th} attribute of x_i
x_{ij}	Short-hand for $x_i.a_j$
Dom_j	Domain of the attribute a_j
$s_j(.,.)$ or $s(.,.)[j]$	Similarity between two values (specified as parameters) of the jth attribute in schema $\mathcal{A}$
$d_j(.,.)$ or $d(.,.)[j]$	Dissimilarity between two values (specified as parameters) of the jth attribute in schema $\mathcal{A}$
q	Query object
$\mathcal{O}$	A similarity search operator
$s(q,x)$	The vector of attribute-wise similarities between the query and the object x
$d(q,x)$	The vector of attribute-wise dissimilarities between the query and the object x
$s_{\mathcal{O}}(q,x)$	The intermediate operator-specific representation of similarity between q and x
$d_{\mathcal{O}}(q,x)$	The intermediate operator-specific representation of dissimilarity between q and x
$R_{\mathcal{O}}(q,\mathcal{X})$	The result set for the query, operator combination on the dataset $\mathcal{X}$

Table 3.1: Notation Table

The system would then compare the query to a data objects in the system and generate a (dis)similarity vector for each comparison; the (dis)similarity vector obtained by comparing q with a data object $x \in \mathcal{X}$ would be as follows:

$$s(q,x) = [s_1(q.a_1,x.a_1), s_2(q.a_2,x.a_2), \ldots, s_m(q.a_m,x.a_m)] \qquad (3.1)$$

If the dissimilarity function is used instead, the corresponding vector would be denoted as $d(q,x)$. It may be noted that estimating the $d(.,.)$ and $s(.,.)$ vectors are done in an operator-agnostic fashion. These vectors would now be condensed into operator-specific (dis)similarity estimates that we will denote as $s_{\mathcal{O}}(q,x)$:

$$s(q,x) \xrightarrow{\mathcal{O}} s_{\mathcal{O}}(q,x) \qquad (3.2)$$

As would be obvious by now, the usage of dissimilarity functions would yield a dissimilarity representation $d_{\mathcal{O}}(q,x)$. We will call these as the operator-specific intermediate (dis)similarity representations. The $s_{\mathcal{O}}(q,x)$ need not necessarily be a vector; in fact, the most popular weighted sum operator aggregates the individual similarities to arrive at a single (scalar) numeric measure of similarity. On the other hand, the corresponding transformation for the skyline operator is the identity function that simply copies the input to the output. To summarize the transformations by these two popular operators:

$$s_{\mathcal{O}}(q,x) = \begin{cases} \sum_{1 \leq j \leq m} w_j \times s_j(q.a_j, x.a_j) & if \ \mathcal{O} = weighted \ sum \\ sim(q,x) & if \ \mathcal{O} = skyline \end{cases} \quad (3.3)$$

There are operators whose transformation do not depend on all values in the $s(q,d)$ vector; for example, the k-n-match operator [30] picks the nth highest numeric value in the $s(q,x)$ vector for its scalar intermediate representation. A naive similarity search system that does not use indexes would build the $s_{\mathcal{O}}(q,x)$ representation for each object x in $\mathcal{X}$. These intermediate representations would be then fed into the operator-specific result set estimation module that would then estimate the result set as $R_{\mathcal{O}}(q, \mathcal{X})$. For example, a range operator that requires the result set to comprise only those objects whose aggregate similarity falls within a pre-specified threshold would simply assess the result set as the following:

$$R_{\mathcal{O}}(q, \mathcal{X}) = \{x | x \in \mathcal{X} \wedge sim_{weighted \ sum}(q,x) \geq t\} \quad (3.4)$$

Figure 3.1 depicts the various phases of the similarity search framework just described through a schematic illustration.

While the above procedural walk through of a naive similarity search system did help in introducing most of the notation that we would need for the remainder of this book, it is appropriate to include a few words of caution. Unlike the preceding narrative, most similarity search systems do not perform a comparison between the query and each object in the database since pre-built operator-specific indexes often aid in narrowing down the search to some subsets of the dataset. Second, the transformation to the intermediate representation and the result set estimation phases need not be done sequentially, and could be interleaved for efficiency reasons. Lastly, some similarity search systems that employ operators that can take in multiple queries do not fit into the above framework. However, we believe that the framework is general enough to encompass most common similarity search operators. Let us now look at these two steps in further detail.

3.2 Aggregation Functions

The first step of a similarity search operator is to condense the (dis)similarity vectors representing the attribute wise (dis)similarity scores:

$$d(q,x) \xrightarrow{\mathcal{O}} d_{\mathcal{O}}(q,x) \quad (3.5)$$

The dimensionality of $d_{\mathcal{O}}(q,x)$ can be either same or lesser than that of $d(q,x)$, depending on the aggregation used. The trivial aggregation method is the $Identity(I)$ operator, that keeps the (dis)similarity vector unchanged, i.e. $d_I(q,x) = d(q,x)$

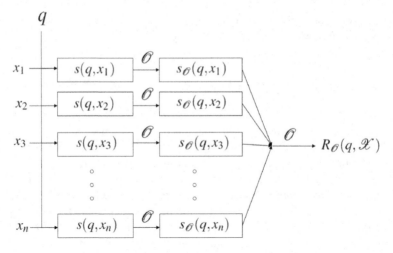

Fig. 3.1: Similarity Search Framework with Notations

where d_I denotes the identity transformation. The *Identity* aggregation is used when there is no meaningful way to combine the (dis)similarity scores across attributes. We will look at the other commonly used aggregation functions next.

3.2.1 Weighted Sum and Euclidean Distance

The weighted sum (WS) function simply computes a weighted sum of the (dis)similarity scores to produce a single score (scalar value) for the object. Given the weights of the attributes in a weight vector, the weighted sum can be specified as:

$$d_{WS}(q,x) = \sum_{1 \le j \le m} w_j \times d(q,x)[j] \tag{3.6}$$

Figure 3.2 shows an example of weighted sum computation for a dataset with two dimensions with weight vector $[1,2]$. The sub-figure on the right shows the locus of points that have the same weighted sum distance when the weights across attributes are identical. The weighted sum function is useful when the distances between different attributes can compensate each other, which enables all the distances to be aggregated into a single score based on the relative importance of the attributes (weights). When the objects are in an Euclidean space, the L_2 or Euclidean distance could be used instead of the weighted sum to compute the aggregate score:

$$d_{L_2}(q,x) = \sqrt{\sum_{1 \le j \le m} d(q,x)[j]^2} \tag{3.7}$$

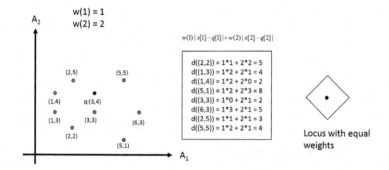

Fig. 3.2: Weighted Sum Computation

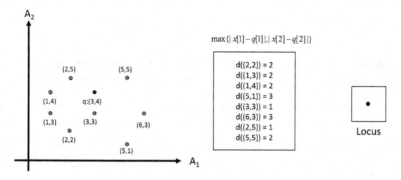

Fig. 3.3: Max Computation

3.2.2 Max

The Max (*max*) function also computes a single scalar score by picking the maximum dissimilarity score among all the attributes.

$$d_{max}(q,x) = \max_{1 \leq j \leq m} d(q,x)[j] \tag{3.8}$$

Figure 3.3 shows an example of the *max* computation for a dataset with two dimensions. It also shows the locus of points that have the same *max* distance from the query *q*. The *max* operator is useful when the user needs to bound the maximum distance between the query and the data object along all dimensions.

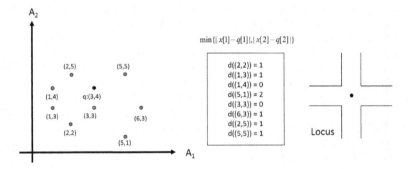

Fig. 3.4: Min Computation

3.2.3 Min

The Min (*min*) function is complementary to max and computes a single scalar score by picking the minimum dissimilarity score among all the attributes.

$$d_{min}(q,x) = \min_{1 \le j \le m} d(q,x)[j] \qquad (3.9)$$

Figure 3.4 shows an example of the *min* computation for a dataset with two dimensions. It also shows the locus of points that have the same *min* distance from the query q. The *min* function is useful when a good match along any one attribute is sufficient for the object to be included in the result set.

3.2.4 N-Match

Instead of considering all the attributes, the N-Match (*NM*) function chooses N attributes that have the best match (least dissimilarity) with the query. It orders these attributes in the increasing order of dissimilarity and chooses the Nth attribute as the final aggregate. It also produces a single value (scalar) for each object.

$$d_{NM}(q,x) = d(q,x)[j]$$

j is the index of the Nth value when elements in $d(q,x)$ are sorted in increasing order
$$(3.10)$$

Figure 3.5 shows an example of the N-match computation for a dataset with two dimensions. The first table shows the attribute wise distances and the second shows the distances ordered for each object. The N-match would return the values in the first column for $N = 1$ and those in the second column for $N = 2$. The N-match

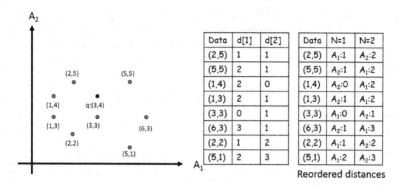

Fig. 3.5: N-Match Computation

function is useful in the scenario where a match on at least N attributes is required for objects to be in the result set.

The N-Match function can be thought of as a generalization of the *min* and *max* functions. N-Match with N set to 1 corresponds to the *min*, since it picks the best matching attribute. Similarly, N-match with N set to the total number of attributes corresponds to the *max*, since it picks the attribute with the largest distance.

3.3 Filter Functions

After condensing the (dis)similarity vectors by aggregation, they are fed into a filter step that chooses a subset of objects to be in the final result set $R_{\mathscr{O}}(q, \mathscr{X})$ based on these (dis)similarity scores. As with other steps, there is choice of various types of filters that can be used. Some of the common ones are described below.

3.3.1 Threshold Filter

This is a simple filter that uses a threshold on each component of the condensed vector $d_{\mathscr{O}}(q,x)$. Objects that satisfy the threshold criteria are included in the final result set. Note that the threshold is a single value (scalar) when $d_{\mathscr{O}}(q,x)$ is a scalar. Given a threshold vector t, the result set is computed as shown below:

$$R_{\mathscr{O}}(q, \mathscr{X}) = \{x | x \in \mathscr{X} \land \forall i, \ d_{\mathscr{O}}(q,x)[i] \leq t[i]\} \qquad (3.11)$$

In the example shown for weighted sum in Figure 3.2, if the threshold were set to 4, the final result set R would be $\{(1,3),(1,4),(3,3)\}$, since all of them have a distance of at most 4 from the query object. The threshold filter is used when one needs to impose upper bounds on the distances of objects in the result set to the query; for example, if the $d_\Theta(q,x)$ is designed to be a scalar denoting the distance between the two objects, the threshold filter translates to choosing all objects within a distance threshold from the query.

3.3.2 Top-k Filter

This filter applies to cases where the condensed vector $d_\Theta(q,x)$ is actually a single value (scalar). The top-k filter ranks the objects based on the aggregated distance and chooses the k objects which have the smallest distance from the query object. The result set $R_\Theta(q, \mathcal{X}) \subseteq \mathcal{X}$ satisfies the following conditions:

$$|R_\Theta(q, \mathcal{X})| = k$$
$$x \in R_\Theta(q, \mathcal{X}) \Leftrightarrow \forall x' \in \mathcal{X} - R_\Theta(q, \mathcal{X}), d_\Theta(q,x) \le d_\Theta(q,x') \tag{3.12}$$

In the example shown for weighted sum in Figure 3.2, if k is 2, the final result set R would be $\{(1,4),(3,3)\}$, since these two objects are nearest to the query object. The top-k filter is used when there is a bound on the number of objects that can be returned in the result set.

3.3.3 Skyline Filter

The skyline filter is applicable when the condensed vector $d_\Theta(q,x)$ has more than one dimension, i.e. when it is not possible to reduce the original (dis)similarity vector $d(q,x)$ into a single scalar value. The skyline consists of a set of objects that are not dominated by other objects. Domination is usually assessed using the (dis)similarity vector $d_\Theta(q,x)$ with reference to a query object. An object dominates another if it is at least as similar to the query object on all dimensions and strictly more similar in at least one dimension. More formally, an object x is said to dominate another object y with respect to a query object q, (represented as $X \succ_Q Y$) iff:

$$\forall i, d_\Theta(q,x)[i] \le d_\Theta(q,x)[i] \; and$$
$$\exists i, d_\Theta(q,x)[i] < d_\Theta(q,x)[i] \tag{3.13}$$

It may be noted that the second condition above ensures that duplicates (i.e., objects which have the same value for all attributes) do not dominate one another. In this setting, for a query q, the skyline filter retains objects from $\mathcal{X}$ that are not dominated (with respect to q) by *any* other object in $\mathcal{X}$. The result set $R_\Theta(q, \mathcal{X}) \subseteq \mathcal{X}$ satisfies the following conditions:

$$\forall x \in R_\Theta(q, \mathcal{X}), \nexists x' \in \mathcal{X} \; such \; that \; x' \succ_q x$$
$$\forall x \in (\mathcal{X} - R_\Theta(q, \mathcal{X})), \exists x' \in \mathcal{X} \; such \; that \; x' \succ_q x \tag{3.14}$$

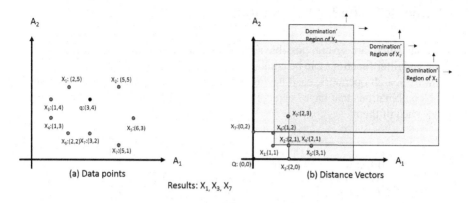

Fig. 3.6: Skyline Filter

The first condition means that there are no objects dominating the ones in $R_\mathcal{O}(q, \mathcal{X})$ whereas the second condition ensures that $R_\mathcal{O}(q, \mathcal{X})$ covers all non-dominated objects. In certain cases where the user may not want to distinguish between duplicate objects, $R_\mathcal{O}(q, \mathcal{X})$ may be pruned to remove the duplicates. It may be observed that $R_\mathcal{O}(q, \mathcal{X}) = \mathcal{X}$ in the worst case, but, in most practical scenarios, $|R_\mathcal{O}(q, \mathcal{X})| << |\mathcal{X}|$. Further, it may be noted that cases were $R_\mathcal{O}(q, \mathcal{X})$ contains most of the objects in $\mathcal{X}$ may not be interesting to the user as she would then have too many results to analyze.

Figure 3.6 shows the skyline computation for a set of data points with two attributes. The final result set R consists of three objects $\{X_1 : (2,5), X_3 : (1,4), X_7 : (3,2)\}$. Figure 3.6(a) shows the original points with the query object, whereas Figure 3.6(b) shows the distance vectors $d_\mathcal{O}(q,x)$ for each point w.r.t. the query q. The regions of domination for each of these objects can be seen from Figure 3.6(b). Note that all the remaining objects lie in the domination region of at least one object in the result set. For example, $X_5 = (6,3)$ is dominated by both $X_1 = (2,5)$ and $X_3 = (1,4)$ with reference to the query $q = (3,4)$. The dominance of X_1 is due to the fact that the distance of X_5 from q on the first attribute is 3, which is greater than the distance (1) of X_1 from q. On the second attribute, both X_1 and X_5 have the same distance (1) from the query q.

The skyline filter is useful when the attribute (dis)similarities cannot be aggregated into a single value and all the attributes are important in deciding the set of similar objects. An interesting property of the skyline is that it contains every object that is closest to the query based on some monotone aggregation function of similarities. In addition, for every point in the skyline, there exists a monotone aggregation function that is maximized at that point. Thus, the skyline does not contain any object that is not the best according to some possible weighting.

3.4 Common Similarity Operators

As we saw, there are several possible aggregation and filter functions. These aggregation and filter functions can be combined in different ways, leading to a variety of similarity search operators, each having a different semantics. Table 3.2 lists some of the combinations that have been commonly studied in literature. We will briefly review each of them.

	Threshold	Top-k	Skyline
Identity	Bounding box		Skyline
Weighted Sum		Weighted Sum Top-k	
Euclidean (L_2)	Circular Range	Distance Top-k	
N-Match	Range	K-N-Match [30]	

Table 3.2: Common Similarity Operators

3.4.1 Threshold based Operators

Threshold based operators use a threshold filter along with some aggregation. The simplest case is the Identity aggregation that leads to the bounding box operator. Identity aggregation means that the attribute distance vector is left unchanged. The threshold on the distance for each attribute leads to a range of possible values for that attribute, depending on the location of the query object. Figure 3.7 shows an example of the ranges produced by the thresholds r_1 and r_2 in the two dimensional case for the query point $(3, 4)$. Such bounding box queries are very useful in a variety of scenarios. For example, one may want to find hotels within a price range of $80 and $100 and with average user rating between 3 and 5.

When a L_2 aggregation is used along with a threshold filter, we get a range query, where the range is a circle with the center at the query point and the radius equal to the threshold value. Any object that lies within the circle has its L_2 distance from the query less than the threshold. This form of aggregation is useful in an Euclidean space where the attributes are the coordinates and the distance between two points is the length of the straight line joining them. Finally, when a N-match aggregation is used, only the distance for the Nth best matching attribute is tested against the threshold. The semantics of such a query depends on the value of N. For example, it could correspond to the *Min* operator for $N = 1$ and the *Max* operator for N is number of attributes, as explained in Section 3.2.4. Figure 3.8 shows examples of range queries with L_2 aggregation and the N-match aggregations for the query point $(3, 4)$ and range threshold r.

Several multi-dimensional index structures, such as R-Trees [18], KD-Tree [3] and Quad-Tree [15] have been designed to efficiently answer range queries.

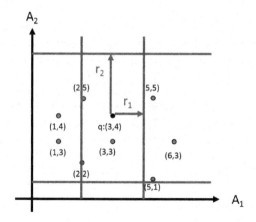

$$Decision\, Criterion : \forall i, |\,(x[i] - q[i])\,| \leq r_i$$

Fig. 3.7: Bounding Box Query

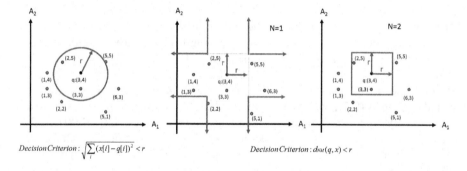

$$Decision\, Criterion : \sqrt{\sum_i (x[i] - q[i])^2} < r \qquad\qquad Decision\, Criterion : d_{NM}(q, x) < r$$

Fig. 3.8: Range Queries

3.4.2 Top-k based Operators

As mentioned in Section 3.3.2, the top-k filter can be combined with aggregate operators that generate a single scalar value from the distance vector. Depending on which aggregate function is used, we get either a Weighted sum top-k, Distance top-k or a K-N-Match operator [30]. Since these are quite straightforward, we will not elaborate on them further. Consider the weighted sum example shown in Figure 3.2. For $k = 2$, the Top-k result would contain the points $(1, 4)$ and $(3, 3)$, since they are the top two points with the least weighted sum distance from the query. If L_2 distance was used instead of the weighted sum, the top-k results would comprise of the points $(3, 3)$ and $(2, 5)$. For the N-Match example shown in Figure 3.5, the top-k results could contain the points $(1,4)$ and $(3,3)$ if $N = 1$. For $N = 2$, there are several possible top-k result sets, including one which would be $\{(3,3),(5,5)\}$.

Top-k retrieval has been extensively studied and many good algorithms and indexing structures have been developed [34]. The most popular simplifying assumption is that the attribute distances satisfy metric properties. This allows the creation of indexing structures that can exploit the triangle inequality property. Examples of metric-space based indexing structures for top-k retrieval include the widely used kd-tree [19] among others [31, 10, 32]. However, the triangle inequality property is too restrictive to model the (dis)similarities as perceived by humans (for example, [17]). This generalized problem is addressed by the family of threshold algorithms [12, 13, 14, 2]. Index based methods for non-metric top-k have also been proposed [9].

3.4.3 Skyline Operators

When the skyline filter is combined with the Identity aggregation, the result is the skyline on the entire set of attributes. This operation is already explained in detail in Section 3.3.3. For example, a skyline on two attributes is shown in Figure 3.6. The Skyline Operator was analyzed in detail for the first time in [4]. Block-Nested-Loops (BNL) and Divide & Conquer were among the very early algorithms proposed for skyline retrieval [4]. Sort-First-Skyline [6] and LESS [16] employ an initial topological sort of objects to reduce the number of comparisons required. Various indexing methods have also been exploited for skyline query processing, including R-trees ([20],[25],[26]), B-trees ([28],[29]) and ALTree [24]. Distributed skyline query processing also has been a subject of recent research ([33],[8]). Middleware algorithms for computing skyline have been first explored by Balke et. al [1]. Numerous variations of the skyline operator and its extensions have also been explored [22, 7, 23, 5, 21, 11, 27].

3.5 Summary

We started this chapter by introducing a simple framework for similarity search systems outlining notations that would be used throughout this book. We then described the common aggregation functions that are used to summarize the (dis)similarity vectors. This was followed by the common filter functions including the threshold, top-k and skyline filters. Finally, we covered the combinations of the aggregation with the filter functions which lead to some of the commonly used similarity search operators. In the next chapter, we will categorize the vast variety of similarity search operators that have been proposed in literature.

References

1. W.-T. Balke, U. Güntzer, and J. X. Zheng. Efficient distributed skylining for web information systems. In *EDBT*, pages 256–273, 2004.
2. H. Bast, D. Majumdar, R. Schenkel, M. Theobald, and G. Weikum. Io-top-k: Index-access optimized top-k query processing. In *VLDB*, pages 475–486, 2006.
3. J. L. Bentley. Multidimensional binary search trees used for associative searching. *Commun. ACM*, 18(9):509–517, 1975.
4. S. Borzsony, D. Kossmann, and K. Stocker. The skyline operator. In *Data Engineering, 2001. Proceedings. 17th International Conference on*, pages 421–430. IEEE, 2001.
5. C.-Y. Chan, H. Jagadish, K.-L. Tan, A. K. Tung, and Z. Zhang. On high dimensional skylines. In *Advances in Database Technology-EDBT 2006*, pages 478–495. Springer, 2006.
6. J. Chomicki, P. Godfrey, J. Gryz, and D. Liang. Skyline with presorting. In *Proceedings of the 19th International Conference on Data Engineering, March 5-8, 2003, Bangalore, India*, pages 717–719, 2003.
7. E. Dellis and B. Seeger. Efficient computation of reverse skyline queries. In *Proceedings of the 33rd international conference on Very large data bases*, pages 291–302. VLDB Endowment, 2007.
8. K. Deng, X. Zhou, and H. T. Shen. Multi-source skyline query processing in road networks, 2007.
9. P. M. Deshpande, P. Deepak, and K. Kummamuru. Efficient online top-k retrieval with arbitrary similarity measures. In *Proceedings of the 11th international conference on Extending database technology: Advances in database technology*, pages 356–367. ACM, 2008.
10. V. Dohnal, C. Gennaro, P. Savino, and P. Zezula. D-index: Distance searching index for metric data sets. *Multimedia Tools Appl.*, 21(1):9–33, 2003.
11. T. Emrich, M. Franzke, N. Mamoulis, M. Renz, and A. Züfle. Geo-social skyline queries. In *Database Systems for Advanced Applications*, pages 77–91. Springer, 2014.
12. R. Fagin. Combining fuzzy information from multiple systems. In *PODS*, pages 216–226, 1996.
13. R. Fagin. Combining fuzzy information: an overview. *SIGMOD Record*, 31(2):109–118, 2002.
14. R. Fagin, A. Lotem, and M. Naor. Optimal aggregation algorithms for middleware. *Journal of Computer and System Sciences*, 66(4):614–656, 2003.
15. R. A. Finkel and J. L. Bentley. Quad trees: A data structure for retrieval on composite keys. *Acta Inf.*, 4:1–9, 1974.
16. P. Godfrey, R. Shipley, and J. Gryz. Maximal vector computation in large data sets. In *Proceedings of the 31st International Conference on Very Large Data Bases, Trondheim, Norway, August 30 - September 2, 2005*, pages 229–240, 2005.
17. K. Goh, B. Li, and E. Chang. Dyndex: A dynamic and nonmetric space indexer, 2002.
18. A. Guttman. R-trees: A dynamic index structure for spatial searching. In *SIGMOD'84, Proceedings of Annual Meeting, Boston, Massachusetts, June 18-21, 1984*, pages 47–57, 1984.
19. I. Kalantari and G. McDonald. A data structure and an algorithm for the nearest point problem. *IEEE Trans. Software Eng.*, 9(5):631–634, 1983.
20. D. Kossmann, F. Ramsak, and S. Rost. Shooting stars in the sky: An online algorithm for skyline queries. In *VLDB*, pages 275–286, 2002.
21. C. Li, N. Zhang, N. Hassan, S. Rajasekaran, and G. Das. On skyline groups. In *Proceedings of the 21st ACM international conference on Information and knowledge management*, pages 2119–2123. ACM, 2012.
22. X. Lian and L. Chen. Monochromatic and bichromatic reverse skyline search over uncertain databases. In *Proceedings of the 2008 ACM SIGMOD international conference on Management of data*, pages 213–226. ACM, 2008.
23. X. Lin, Y. Yuan, Q. Zhang, and Y. Zhang. Selecting stars: The k most representative skyline operator. In *Data Engineering, 2007. ICDE 2007. IEEE 23rd International Conference on*, pages 86–95. IEEE, 2007.

24. D. Padmanabhan, P. M. Deshpande, D. Majumdar, and R. Krishnapuram. Efficient skyline retrieval with arbitrary similarity measures. In *EDBT*, pages 1052–1063, 2009.

25. D. Papadias, Y. Tao, G. Fu, and B. Seeger. An optimal and progressive algorithm for skyline queries. In *SIGMOD Conference*, pages 467–478, 2003.

26. D. Papadias, Y. Tao, G. Fu, and B. Seeger. Progressive skyline computation in database systems. *ACM Trans. Database Syst.*, 30(1):41–82, 2005.

27. M. Sharifzadeh and C. Shahabi. The spatial skyline queries. In *Proceedings of the 32nd international conference on Very large data bases*, pages 751–762. VLDB Endowment, 2006.

28. K.-L. Tan, P.-K. Eng, and B. C. Ooi. Efficient progressive skyline computation. In *VLDB*, pages 301–310, 2001.

29. Y. Tao, X. Xiao, and J. Pei. Subsky: Efficient computation of skylines in subspaces. In *ICDE*, page 65, 2006.

30. A. K. Tung, R. Zhang, N. Koudas, and B. C. Ooi. Similarity search: a matching based approach. In *Proceedings of the 32nd international conference on Very large data bases*, pages 631–642. VLDB Endowment, 2006.

31. J. K. Uhlmann. Satisfying general proximity/similarity queries with metric trees. *Information Processing Letters*, 40(4):175–179, 1991.

32. E. Vidal. New formulation and improvements of the nearest-neighbour approximating and eliminating search algorithm (aesa). *Pattern Recognition Letters*, 15(1):1–7, 1994.

33. S. Wang, B. C. Ooi, A. K. H. Tung, and L. Xu. Efficient skyline query processing on peer-to-peer networks. In *ICDE*, pages 1126–1135, 2007.

34. P. Zesula, G. Amato, V. Dohnal, and M. Batko. *Similarity Search - The Metric Space Approach*. Springer, 1978.

Chapter 4
Categorizing Operators

A common assimilation aid that is used in surveys is to organize the various methods in the field in the form of a hierarchy with each level in the hierarchy dealing with a particular choice made by the techniques. Using such a taxonomy for similarity operators would yield two levels, the first describing the type of the summary representation and the second specifying the way the result set may be computed; we have discussed these phases extensively in the first two chapters. Despite being conceptually distinct, most similarity operator implementations tend to interleave these phases for efficiency considerations, thus making the two-phase narrative not so useful from a practical standpoint. Thus, we will organize similarity operators using a more utilitarian classification into four types. Later, we will introduce *features*, a set of add-on functionalities that could be used to enrich the specification of a similarity operator, to enhance applicability to specific scenarios.

4.1 Types of Similarity Operators

We now classify similarity operators based on two considerations as follows:

- **Ordered vs. Unordered Result Set:** Some similarity operators enforce an ordering of objects in the result set whereas some others do not. The oft used weighted sum top-k operator uses ordering in the result output where the object most similar to the query is placed at output slot #1, followed by the next most similar object and so on. On the other hand, the skyline operator does not specify an ordering of the result objects and simply outputs a subset of objects that it identifies, as the result set.
- **Role of Attributes in Result Set Determination:** Certain operators require information of all the attributes of an object, to fully determine its status in the result set. For example, the weighted sum top-k operator scores objects based on the aggregate of the similarity to the query across attributes, and chooses the top-

© The Author(s) 2015
D.P and P.M. Deshpande, *Operators for Similarity Search*,
SpringerBriefs in Computer Science, DOI 10.1007/978-3-319-21257-9_4

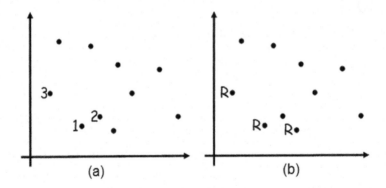

Fig. 4.1: Ordered vs. Unordered Output

k objects[1]. In the case of the K-N-match [21] operator, the score of each object is the similarity on the Nth most similar attribute, and when N is set to 1, the score of an object is fully determined by the similarity on the most similar attribute. Thus, these scenarios represent opposite ends of the spectrum in that the former requires information of all attributes of an object to determine its status in the result set, whereas the K-1-match operator can determine the score of an object as soon as it is known that the attribute on which the object has maximum similarity to the query (e.g., as estimated using bounds, as is usually done) has been seen.

It may be noted that these classifications are somewhat orthogonal to the classification based on *aggregation* and *filter* functions as we saw in the previous chapter. We will now delve a little deeper into the above two criteria, and outline the implications of the choices available for each. We will also analyze some example operators in this 4-class type classification formed by the combination of two choices on the two criteria above.

4.1.1 Ordered vs. Unordered Result Set

Consider the two scenarios outlined in Figure 4.1(a) and (b). In both of these, the query is assumed to be at the origin, and the x and y co-ordinates of each object is determined according to the distance of the object from the query on the x and y attributes; in this toy example, we will assume that the schema contains just two attributes. Figure 4.1(a) illustrates the result set of a weighted sum top-k operator

---

[1] There are techniques (e.g., [5]) that exploit bounds in similarity values to bound the similarity of an object to the query, and can determine the membership in the top-k result set using such bounds; however, these would be unable to provide the actual score of each object in the result, and are hence not applicable in scenarios that need quantifying the weighted sum similarity exactly.

that uses uniform weights, with the result slots marked against the respective data objects. Thus, the closest data object is marked with slot #1, whereas the next closest object is marked with slot #2. On the other hand, in the case of the skyline operator in Figure 4.1(b), there is no ordering among the result objects marked with R. It may be noted that it is not even possible to enforce an artificial ordering based on the distance to the query object since a linear ordering requires an aggregation of similarities across attributes to a single value (per object) which is not possible when the method of aggregation and the relative importance of attributes (such as weights) are not specified in the operator. Thus, this criterion leads to two classes of operators as below.

Ordered Operator: An operator whose semantics specify an ordering of the objects in the result.
Unordered Operator: An operator whose result is an unordered set.

Implications: While being a relatively simple distinction conceptually, the presence or absence of ordering among result objects has several implications on design of algorithms to implement similarity operators. Consider an online algorithm (e.g., [3]) for similarity search that emits results one by one as and when they are identified. Such an algorithm that works with a similarity operator producing ordered output would need to wait until the ordering information becomes available; i.e., it is not just enough to quantify the weighted sum of similarities of a particular object from the query in the weighted sum top-k operator since the rank of the object in the output is determined by the number of objects that exist in the dataset with a higher similarity value to the query. On the other hand, the unordered skyline operator can emit a an object as soon as it is identified that no other objects in the dataset dominate it. Similarly, membership in the result set of an unordered range query is determined by whether the object falls within the specified range from the query, and does not depend on the distance of other objects from the query. Ordered similarity operators are advantageous to use when results are to be delivered on interfaces where resource constraints such as the available screen size vary. This is so since the ordering among the results can be leveraged to pick the *first k* results for any arbitrary value of k. Many unordered similarity operators such as the skyline or range operators can produce result set of varying cardinality, and it is not intuitive to select a subset of k objects to show on a constrained interface such as a mobile device screen.

4.1.2 All vs. Some: Role of Attributes in Result Set Determination

We will now revisit the Similarity Search framework introduced in Chapter 2, the schematic diagram of which is reproduced in Figure 4.2. As may be recollected, the $s(q,x)$ notation represents a vector of length equal to the number of attributes, m. Each element of the $s(q,x)$ vector denotes the similarity of the object x to the

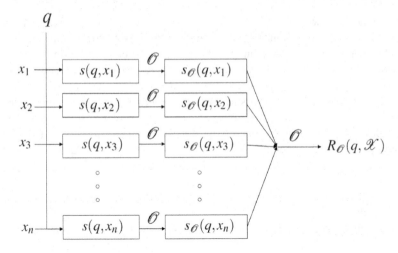

Fig. 4.2: Similarity Search Framework

query on a specific attribute. We will now pay attention to the operator specific transformation of this vector:

$$s(q,x) \xrightarrow{\mathcal{O}} s_{\mathcal{O}}(q,x) \qquad (4.1)$$

The key aspect of interest for our second criteria to classify similarity operators is as to how many elements of the $s(q,x)$ vector *directly* contribute to the computation of the $s_{\mathcal{O}}(q,x)$ vector. To illustrate what we mean by *direct* contribution, let us look at the example of the K-N-match operator [21], where objects are ranked with respect to their similarity to the query on the Nth most similar attribute. For $N = 1$, we would simply choose the similarity on the most similar attribute to represent the object-wise similarity. As an example, suppose the query object is a fruit, say *apple*, being posed against the dataset of fruits. An *orange*, which bears a high similarity to the query on the shape attribute, would be regarded as highly similar to the query on the K-1-match operator regardless of the apparent dissimilarities on other attributes such as *taste*, *color* and *fruit-family*. The intermediate representation for the K-1-match operator is denoted by the following transformation:

$$s_{\mathcal{O}}(q,x) = max\{s(q,x)[i] | 1 \le i \le m\} \qquad (4.2)$$

Under the above transformation, for our apple-orange example, we will regard only the *shape* attribute to have directly contributed to the transformed representation. It may be noted that the other attributes may be thought of as contributing indirectly, since their exclusion is due to them having a lower value than the similarity on the shape attribute. Further, when the fruit *pear* is compared against the query, the direct contribution could be from the *fruit-family* attribute since *apple* and *pear* belong to very similar fruit-families. Thus, this notion of direct contribution is object specific.

Towards classifying similarity operators, we are not interested in the specific attribute that directly contributes to the transformed representation, but simply as to whether *all* attributes contribute to the transformed representation. Unlike the K-1-match example above, similarities on all the attributes are seen to contribute to the transformation in the weighted sum operator:

$$s_\mathcal{O}(q,x) = \sum_{i=1}^{m} w_i \times s(q,x)[i] \tag{4.3}$$

In the transformation for the skyline operator, where the $s_\mathcal{O}(q,x)$ is simply the $s(q,x)$ vector itself, *all* attributes are seen to be directly contributing since the similarities on each attribute are carried forward to the transformed representation. Without much ado, the classification under this criterion is as outlined below.

All-Attribute Operator: An operator where similarities on all attributes directly contribute to the transformed operator-specific representation.

Some-Attribute Operator: An operator where only the similarities on some attributes directly contribute to the transformed operator-specific representation.

Implications: The criteria as to whether a similarity operator uses direct contribution from all attributes or only some has implications in the design of indexes and algorithms that work on such indexes. Consider the case of an operator such as r-attribute range search whose semantics is such that an object would be included in the result set if it satisfies a range constraint on any set of r attributes (where r is pre-specified); this is clearly a some-attribute operator since a subset of r attributes determine the membership of an object in the result set. In this case, a simple method to parallelize the processing would be to build multiple indexes on different combinations of r attributes, and fire the search query on all of them. An object that satisfies the distance constraint on *any* of the individual indexes would qualify to be in the result since each index works with a legitimate combination of r attributes. Thus, partitioning objects based on attributes, as is done in the case of column stores[2], could be more appropriate to some-attribute operators than for all-attribute operators. In an all-attribute operator such as weighted sum, upper and lower bounds on the similarities on specific attributes may be used as tools to prune objects from consideration even without seeing all their attributes. As an example, consider the case of attribute similarities in the $[0,1]$ range and an object whose aggregate similarity on $m-1$ attributes is s_x. Since there is only one more attribute on which similarity needs to be assessed, the total similarity would fall within the range $[s_x, s_x + 1.0]$ assuming a weight of unity for the attribute yet to be seen. Such bounds may be used to prune objects as has been explored in the threshold family of algorithms [5]. Though not a general principle, algorithms for all-attribute operators tend to work with bound-based pruning strategies whereas algorithms for some-attribute

---

[2] http://en.wikipedia.org/wiki/List_of_column-oriented_DBMSes - Accessed February 19th, 2015

operators try to organize indexes so as to traverse the direct contributor attributes up front.

	Ordered	Unordered
All-Attribute	Weighted Sum Top-k	Skyline
	Reverse k-Ranks [25]	RkNN [18]
Some-Attribute	K-N-Match [21]	Subspace Range Query [11]
	Subspace Top-k [11]	

Table 4.1: Examples of Similarity Operators Classification

4.1.3 Example Operators under the 4-type Classification

Table 4.1 summarizes the four combinations in the classification according to the two criteria discussed so far. We include some example operators in the table to illustrate that there have been operators proposed in the literature that fall into each of the four types. We will use these four type classification as a tool to position each of the operators that we will introduce in subsequent chapters. We conclude this section with a brief discussion on each of the four types of similarity operators.

Ordered All-Attribute Operators: These make use of all the attributes in scoring a data object, and typically project it to a 1-dimensional space by using a scalar aggregate score such as a weighted sum. Thus, the transformed representation in ordered operators (i.e., $s_{\mathcal{O}}(q,x)$) is usually a scalar. For the Reverse-k-Ranks query [25], the scalar score associated with each data object is the rank of the query object in a weighted sum query centered on the object (hence the adjective reverse in the name of the operator). The 1-dimensional ordering naturally yields an ordered output.

Ordered Some-Attribute Operators: Operators in this category differ from the above in not utilizing the similarities on all the attributes in computing the total ordering by means of a scalar score. Thus, these could use the similarity on the attribute with the highest similarity (i.e., K-1-match), or the Nth highest similarity (i.e., K-N-match), or the highest dissimilarity (e.g., L_p-norm with $p \rightarrow \infty$). Alternatively, subspace operators such as Subspace Top-k [11] use the weighted sum aggregation on a subset of attributes in the schema. Once the total order is established by means of a scalar scoring function, the ordering in the output is trivially defined.

Unordered All-Attribute Operators: Most operators under this category do not associate a single scalar score for each object; exceptions do exist, such as the range query where a weighted sum-type score is used. The usual method is to use the similarity vector as an input to an operator-specific operation that determines the membership of the object in the result set. For example, the skyline operator checks for the presence of a dominating object, and determines the result set membership accordingly. The RkNN operator [18] checks for the membership of the query object in the set of k nearest neighbors of the data object as the criterion to determine result set membership. The objects identified to be in the result set would then be output as an unordered set.

Unordered Some-Attribute Operators: Our last type of operators, as may be understood by the definition, fall into the category of determining membership of objects using their similarity vectors, without using all the attributes. Any operator in the unordered all-attribute category could have a companion operator in this category if the membership function needs to be enforced only on a subset of attributes. The subspace range query exactly does this, as discussed in Section 4.1.2.

4.2 Features: Add-on Functionalities for Similarity Operators

We will now discuss some tools to enhance the semantic richness of operators, in order to tune them to scenarios that warrant such enhancements. We call these as add-on functionalities, or features, since they can be added on to most basic operators to create richer ones. That said, some features are incompatible with some operators, and some features are incompatible with certain others - so, they can't be both used together. We will now outline some features that have been widely explored within the similarity search community.

4.2.1 Indirection: Forming "Reverse" Operators

We will first introduce the indirection feature somewhat formally, and then proceed to an example based illustration outlining the motivation and example usages of the feature. Consider an operator $\mathcal{O}$ whose result set on a dataset $\mathcal{X}$ be denoted as $R_{\mathcal{O}}(q, \mathcal{X})$. Now, the result set of the operator $\mathcal{O}$ enhanced with the indirection feature - we will denote this enhanced operator as $\mathcal{O}^{R}$ [3] - would be specified as:

$$R_{\mathcal{O}^R}(q, \mathcal{X}) = \{x | x \in \mathcal{X} \wedge q \in R_{\mathcal{O}}(x, \mathcal{X} \cup \{q\} - \{x\})\} \qquad (4.4)$$

where $R_{\mathcal{O}}(x, \mathcal{X} \cup \{q\} - \{x\})$ denotes the result set for the operator $\mathcal{O}$ when x is used as a query against the dataset where x is replaced by the original query q. It may be easier to understand the semantics of the indirection feature by outlining a naive 2-step implementation of the indirection operator as follows:

- For each object x in the dataset $\mathcal{X}$, run the $\mathcal{O}$ operator using x as the query against the dataset with x replaced by q. Collect the resulting output $R_{\mathcal{O}}(x, \mathcal{X} \cup \{q\} - \{x\})$, in each case.

---

[3] Typically, operators that use the indirection feature are called *reverse* operators as we will see shortly; the R in the $\mathcal{O}^R$ is chosen to denote the *reverse* operator.

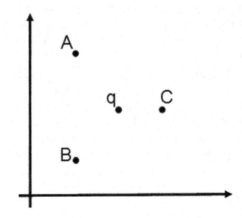

Fig. 4.3: Indirection on Nearest Neighbor (Top-1) Operator: The result of the NN operator with q as query yields C, whereas the reverse NN operator yields a result set comprising A, B and C

- Sift through the various $R_{\mathcal{O}}(x, \mathcal{X} \cup \{q\} - \{x\})$s, and determine those xs for which the corresponding result set from the first phase contain q. Output those objects as the resultset $R_{\mathcal{O}R}(q, \mathcal{X})$.

Consider the weighted sum top-k operator whose result set is formed by the k objects in $\mathcal{X}$ that are most similar to the query based on the weighted sum aggregation of similarities across attributes. The corresponding reverse operator would then comprise objects such that the query is among the k objects closest to them. While the top-k operator has a result set cardinality bounded by k, it is not necessarily true for the reverse top-k operator. Figure 4.3 illustrates this by means of an example on the Top-1 operator (also sometimes called nearest neighbor operator) in a 2-dimensional euclidean space; the nearest neighbor of q is just the set $\{C\}$, whereas the Reverse NN of q produces a result set containing the entire dataset, i.e., $\{A, B, C\}$. This is so since q is the closest object to each of the three objects in the dataset, as may be easily observed visually.

Motivation: The most common motivation that has been cited for the indirection feature is to use indirection as a tool to flip the perspective from that of the consumer to that of a service provider. Consider an entrepreneur planning to setup a food joint in one of multiple possible locations; she wants to prioritize candidate locations using a database of the residence location of each resident in the locality, as well as the location of other food joints in the city. If the entrepreneur is willing to go by a simplistic assumption that a consumer would typically choose one among the top-10 locations closest to him/her to dine in, the potential user base for each location may simply be estimated as the cardinality of the Reverse top-10 set for each location. The Reverse top-10 set for a location identifies consumers who have that location among the top-10 closest locations to dine in. As is obvious, the user should be careful to steer the search so that only food joints (and not other consumer locations) are considered for the reverse top-10 query. The raw top-10 operator is a tool for the consumer to choose the service provider to avail services from. Thus, the indirection feature is the analogous tool for service providers to decide on desired attributes for a proposed/existing service. Other applications of the indirection operator include determining the *impact* of a product compared to competitors' products, and identifying a set of potential customers to send targeted offers to. Yet another scenario where indirection is useful is that related to resiliency planning; we will describe that in detail while introducing specific operators.

Basic Operator	Operator with Indirection Feature Added
Nearest Neighbor	Reverse Nearest Neighbor [8]
Farthest Neighbor	Reverse Farthest Neighbor [8]
Top-k	RkNN [18]
Skyline	Reverse Skyline [2]
k Farthest Neighbors	Reverse k Farthest Neighbor [9]
Visible Nearest Neighbors	Visible RkNN [6]
Skyband [15]	Reverse k-Skyband [13]

Table 4.2: Examples of Operators with Indirection Feature

Examples: Table 4.2 lists a few examples of operators where the indirection feature is used. We will see many of these operators in some amount of detail in a later chapter.

4.2.2 Chromaticity: Channelizing the Search to Specific Types

A dataset could legitimately be comprised of multiple types of data; chromaticity is a feature that restricts the result set to contain data objects based on specific types. We will now look at two ways of using (or not using) the types of data in a dataset, as has been explored in literature.

Monochromaticity: Monochromaticity is a feature that may be considered to be used in all operators implicitly. Monochromaticity assumes that all the objects in the dataset (as well as the query) are *colored* with a single type. Thus, searching for the most similar social media profiles - i.e., the top-k operator - using a social media profile as a query implicitly uses the monochromaticity feature. Thus, the result set while using this feature remains the same as $R_{\theta}(q, \mathcal{X})$.

Bichromaticity: Bichromaticity is a feature that requires that the dataset $\mathcal{X}$ be classified into two sets depending on the types of objects contained within it. To understand this better, it is useful to go back to the motivation of the indirection feature in Section 4.2.1; an entrepreneur trying to decide on a location for a planned service issues the *candidate location* as the query, whereas the actual search is performed on the dataset of *consumer locations*. In this example, the two types correspond to *service locations* and *consumer locations*. Bichromaticity feature modifies the operator such that the search for results be done only on the subset of data objects that are **not** of the query type. Let us now look at a formal specification of the result set of $\mathcal{O}^B$, which is the shorthand to denote the $\mathcal{O}$ enhanced with the bichromaticity feature:

$$R_{\mathcal{O}^B}(q, \mathcal{X}) = R_{\mathcal{O}}(q, \{x | x \in \mathcal{X} \wedge x.type \neq q.type\}) \qquad (4.5)$$

where $\{x | x \in \mathcal{X} \wedge x.type \neq q.type\}$ denotes the subset of $\mathcal{X}$ that contains all objects whose type does not match the type of the query. Thus, the result of the bichromatic operator is simply the result of the basic operator when applied to the specified subset. In our example, the bichromatic top-k operator, would return the closest service provider locations when queried with a consumer location, and vice versa. It should be noted that data of both types are expected to come from the same schema; else, the result set for the bichromatic operator may not be well defined. If the query is purely geo-centric as in our example, the service locations as well as consumer locations both come from the same trivial schema that has attributes such as the latitude and longitude of the location being represented. Even in other cases such as food joints, the consumer profile could have attributes such as cuisine preferences and rating preferences that could be easily mapped to the restaurant cuisine specialization and restaurant ratings respectively. The terms monochromatic and bichromatic has been used with different semantics in efforts such as [22]; our discussion excludes those and pertains only to the data type interpretation of chromaticity as outlined so far.

Motivation: The scenario of service providers and consumers provides a compelling motivation for the bichromaticity feature. Thus, in a dataset that has both consumer profiles and service provider profiles, a proximity query (such as top-k, range etc.) issued with the consumer profile would find similar service provider profiles. Similarly, when queried with the service provider profile, the simple proximity query would find similar consumer profiles. However, as we saw in our earlier example, simply ensuring a good match with a consumer profile is not usually sufficient to attract the consumer to the service provider; this is so since such an estimate is agnostic to whether there are other service providers that are even better suited to such consumers. Thus, from the service provider perspective, it is useful to combine the bichromaticity feature with the indirection feature since that would enable finding consumers for whom the service provider is among the most similar service providers.

Examples: Table 4.3 lists some examples of operators that use the bichromaticity feature. As outlined in the motivation, the bichromaticity feature has been mostly explored in combination with the indirection feature.

Basic Operator	Operator with Bichromaticity (and any other features)
Skyline	Bichromatic Probabilistic Reverse Skyline [10]
Nearest Neighbor	Bichromatic Reverse Nearest Neighbor [20]
Top-k	Bichromatic RkNN [23]

Table 4.3: Examples of Operators with Bichromaticity

4.2.3 Visibility: Is there a line of sight?

Consider similarity search on a geographic space such as the latitude-longitude space where distance between them is estimated using measures such as geodesic distance. The visibility feature enables the similarity search system to restrict the similarity search to such objects that have a line-of-sight to the query point. The semantics of the result set of the visbility enhanced operator may be specified as follows:

$$R_{\mathcal{O}^V}(q, \mathcal{X}) = R_{\mathcal{O}}(q, \{x | x \in \mathcal{X} \wedge visible(q,x)\}) \tag{4.6}$$

where $\{x | x \in \mathcal{X} \wedge visible(q,x)\}$ denotes the subset of objects in $\mathcal{X}$ that are *visible* to q. Visibility is defined using a set of obstacles *Obst*, as follows:

$$visible(q,x) = \begin{cases} false & if \ \exists o \in Obst, intersects(o, line(q,x)) \\ true & otherwise \end{cases} \tag{4.7}$$

Informally, q is regarded to be visible to x (or vice versa, since visibility is symmetric) if there does not exist an obstacle in *Obst* that intersects with the straight line joining q and x. Figure 4.4 illustrates an example of a rod-shaped obstacle that intersects with the straight line between the pairs of $[q,C]$ and $[q,D]$. Thus, the operator with the visibility feature on this obstacle set will limit its operation to the two-element dataset $\{A,B\}$. Obstacles need not be rod-shaped or in the form of a straight line; they could be triangles, rectangles, or in general polygonal since the *intersects*(.,.) operation is well defined for any such general shape.

Motivation: A company that wants to maximize the visibility of a billboard would intuitively want to place it at a location where it is visible to a large number of potential consumers. However, in the presence of high-rise buildings and other obstacles, a simple search for number of people within a specific radius of the billboard - perhaps, enhanced using additional filters such as restricting the search to the direction in which the billboard is facing - is likely to even include people whose view of the billboard is obstructed by obstacles. The visibility feature comes in handy in such scenarios and could be readily applied to get a more realistic estimate of the visibility of the billboard. As may be inferred from the ongoing discussion, the visibility feature is mostly applicable in scenarios where the attribute space being considered is a space where the concept of visibility is meaningful. For example, in a hotel database where attributes include things such as *price* and *rating*, straight lines be-

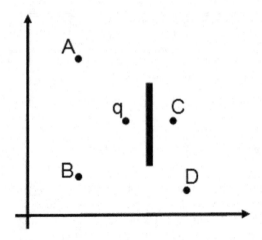

Fig. 4.4: Visbility Example: The obstacle denoted by the black line is seen to intersect the line of sight for the pairs $[q,C]$ and $[q,D]$

tween objects do not map to any semantically useful notion for visbility to make sense.

Basic Operator	Operator with Visibility (and any other features)
Nearest Neighbor	Visible Nearest Neighbor [14]
Top-k	Visible k Nearest Neighbor [14]
Top-k	Visible Reverse k Nearest Neighbor [6]

Table 4.4: Examples of Operators with Visibility

Examples: Due to the limited applicability of the visibility feature, there have only been a few operators in literature that use visibility. Some examples are listed in Table 4.4. It may also be noted that if obstacles are included as points in the dataset, the result of the skyline query would automatically exclude any object that is not visible, from the result set.

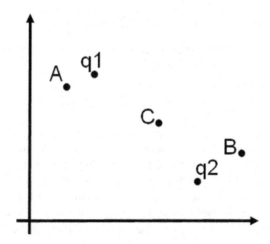

Fig. 4.5: Multi Query Example: While A is the nearest neighbor of $q1$ and B is the nearest neighbor of $q2$, C is seen to be the nearest neighbor for the multi-query $\{q1, q2\}$ when the distance to a multi-query set is determined using the sum of the distances to the individual queries.

4.2.4 Multiple Queries

Many real-world scenarios warrant usage of multiple queries in a similarity search system. In such a scenario, the result set is to be determined with respect to all the queries in the query set. Let the set of queries be denoted as $Q = \{q_1, q_2, \ldots\}$, and the corresponding result set be denoted as $R_{\mathcal{O}M}(Q, \mathcal{X})$.

In general, the result set $R_{\mathcal{O}M}(Q, \mathcal{X})$ of the multi-query operator is not necessarily a function of the result sets obtained by applying the same operator on the individual queries in the query set (i.e., $R_{\mathcal{O}M}(q, \mathcal{X}), q \in Q$). We will illustrate this by means of an example; let us consider a simple multi-query version of the nearest neighbor query where the nearest neighbor is calculated as the object that has the lowest sum of distances across queries. The example in Figure 4.5 illustrates a case where the nearest neighbor for the multi-query $\{q1, q2\}$ is not an element of the nearest neighbor result set of the individual queries. Since C is almost in the middle of the two queries, its aggregate distance to the multi-query is very close to the distance between the two queries itself; this may be easily inferred to be much lower than the aggregate distance to the multi-query from either A or B, the result objects for the individual queries.

Since we cannot easily represent the multi-query by means of other individual query result sets, we will go back to specifying the semantics of the multi-query using the $s(.,.)$ vectors. Let the $s(q, x)$ vector, as we have seen earlier, represent the vector of attribute-level similarities of the data object x to any $q \in Q$. Multi-queries may be used in the transformation to the operator specific representation as follows:

$$\{s(q,x)|q \in Q\} \xrightarrow{\mathcal{O}^M} s_{\mathcal{O}M}(Q,x) \tag{4.8}$$

This transformation for the multi-query nearest neighbor example as illustrated in the example would be the following:

$$s_{\mathcal{O}M}(Q,x) = \sum_{q \in Q} \sum_{1 \leq i \leq m} s(q,x)[i] \tag{4.9}$$

This will be followed by an operator-specific result set estimation step as in the case of the framework outlined in Figure 4.2; the result set estimation is done as would be usually done for the operator $\mathcal{O}$ without the multi-query feature.

Specification of a Multi-Query Feature: Given the intricacies in composing a result set for an operator that uses the multi-query feature, we will now crisply outline what it takes to specify a multi-query feature, for usage in an operator. The core specification is that of a function $f_M(.)$ that does the following transformation:

$$s_{\mathcal{O}M}(Q,x) = f_M(s(q_1,x),s(q_2,x),\ldots) \tag{4.10}$$

The format of the transformed representation, i.e. $s_{\mathcal{O}M}(Q,x)$, needs to be exactly the same as the transformed representation for the raw operator $\mathcal{O}$, i.e., $s_{\mathcal{O}}(q,x)$. After the transformation using $f(\ldots)$, the result set estimation takes on, which is performed exactly as in the case of the raw operator $\mathcal{O}$. If the queries within the set Q are to be weighted differently in the computation of the transformed representation, an additional weight vector $[w_1, w_2, \ldots]$ may be used by the function $f_M(.)$ in doing the transformation.

Motivation: Consider the various social media profiles that people maintain; these could include a *Facebook*[4] profile that lists political and literary interests, a *LinkedIn*[5] profile that lists the work profile including employment history and a *ResearchGate*[6] profile that lists research interests. For a user who is searching for events to attend, such profiles could be used as multiple queries since an event could be considered interesting if it is close enough to any of the various profiles. [17] outlines a spatial scenario where a location may be considered interesting if it is close enough to one of the various locations frequented by the user such as *home*, *office* or *gym*, each of which form queries in the multi-query set.

Examples: Though intuitively useful and appealing, the multi-query feature has not been explored widely in similarity search systems. Among the notable examples where it has been explored includes the multi-query k nearest neighbor operator and algorithm proposed in [17]. This explores an extension of the top-k operator by using the multi-query feature, and also allows for weighting of the different query points in the multi-query set. Another example of usage of multiple queries in similarity search is the spatial skyline operator [16]. Here, a vector of distances is computed for each object where each element in the vector corresponds to the distance

---

[4] https://www.facebook.com/ - Accessed February 21, 2015

[5] http://www.linkedin.com/ - Accessed February 21, 2015

[6] https://www.researchgate.net/ - Accessed February 21, 2015

to a query, followed by a regular skyline computation over such vectors. Note that this does not fit in the standard format for multi-query computation specified above. We will cover this operator in detail in Section 5.3.2.

4.2.5 Subspaces: When Some Attributes are Enough

The subspace feature allows the user of a similarity search system to do the search on only a subset of the attributes that the dataset is defined on. Thus, a subset of attributes are expected to be specified, either explicitly or implicitly, with the query. We will now define the result set of $\mathcal{O}^S$, the operator $\mathcal{O}$ augmented with the subspaces feature, as follows:

$$R_{\mathcal{O}^S}(q, \mathcal{X}) = R_{\mathcal{O}}(q, \Pi_{subspace}(\mathcal{X})) \qquad (4.11)$$

where $\Pi_{subspace}(\mathcal{X})$ denotes the *projection* of the dataset $\mathcal{X}$ on the attributes specified in the set *subspace*. Projection is a simple relational algebra operation that discards the values of the objects in $\mathcal{X}$ on attributes that are not contained in *subspace*. Thus, if the subspace attributes specified are *height* and *weight*, the subspace skyline operator will identify the skyline on the 2-dimensional space composed of *height* and *weight*.

Motivation: In many cases, the data model used by the similarity search system may be too rich for the use case at hand. Having liked a recent movie, one may want to search for similar movies on the attributes {*genre, rating*} without being bothered about other attributes such as {*year of release, director, actors,...*} since the user may not care about such attributes. Using the subspace feature provides a way for the user to tell the system that she considers the other attributes as *don't care* and that the similarity search system should not make use of such attributes in determining the result set. As another example, a user may want to run a skyline search on only a subset of the spatial attributes in the dataset, since a skyline incorporating other dimensions may not make much semantic sense.

Basic Operator	Operator with Subspaces (and any other features)
Nearest Neighbor	Subspace Nearest Neighbor [24]
Nearest Neighbor	Subspace RNN [24]
Range Query	Subspace Range Search [11]
Skyline	Subspace Skyline [19]
Top-k	Subspace Top-k [19]

Table 4.5: Examples of Operators with Subspaces

Examples: As is the case with the multi-query feature, subspaces have also not been explored widely in similarity search systems. However, most of the basic operators have been extended with the subspace feature as can be seen from Table 4.5.

4.2.6 Diversity: To avoid Monotony

Diversity is the feature that caters to the need to prevent an operator from including very similar objects in the result set. This, feature, apart from being like the multi-query feature in not being easily expressible as a function of the result set of the original operator, is not easily composable with any operator as we will see shortly. We will look at the criterion used to ensure diversity in one of the early papers using diversity in retrieval [7]; the result set $R_{\mathscr{O}D}(q, \mathscr{X})$ that denotes the result set of the operator $\mathscr{O}$ is expected to satisfy the following condition:

$$\forall r_i, r_j \in R_{\mathscr{O}D}(q, \mathscr{X}), r_i \neq r_j, divdist(r_i, r_j, divatts) \geq mindiv \qquad (4.12)$$

Informally, this says that for every pair of distinct objects in $R_{\mathscr{O}D}(q, \mathscr{X})$ and a set of diversity attributes *divatts*, the function $divdist(r_i, r_j, divatts)$ should evaluate to at least *mindiv*. Coming to the semantics, $divdist(r_i, r_j, divatts)$ is a function that determines the distance between r_i and r_j on the subset of attributes in *divatts*; the condition ensures that this distance be greater than a threshold specified by *mindiv*. The set of diversity attributes would be a subset of the schema $\mathscr{A}$ and would include attributes on which diversity is to be ensured. The specific implementation of $divdist(.,.)$ in [7] uses a weighted sum of distances between the objects on each attribute in *divatts* with the weight of an attribute determined as being directly related to the quantum of the distance between them on that attribute; this construction helps exaggerate large attribute-specific distances as is deemed to be useful according to the arguments outlined in [7].

Implementing Diversity in Ordered Operators: Since the diversity feature specification does not yield a constructive definition, we will now remark on how it could be implemented in ordered operators, i.e., operators that produce an ordered result set. Ordered operators often use a top-k filter at the end, so that the number of objects to be included in the result set be limited to k. In such a case, a greedy way to produce a diverse output would be to use the following method:

- *Results* $= \phi$
- while($|Results| < k$)

  - Consider the next candidate object in the ordering induced by $\mathscr{O}$
  - Check whether it satisfies the $divdist(.,.,.)$ constraint w.r.t each object already in *Results*
  - If yes, add it to *Results*

Figure 4.6 illustrates an example of the operation of the greedy algorithm based diversity feature on a 2-dimensional Euclidean space. The raw top-3 operator (that

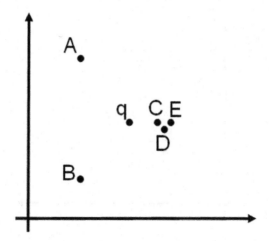

Fig. 4.6: Illustration of the Diversity Feature: In the raw top-3 operator, the result set for the query
q is likely to be $\{C,D,E\}$. However, when the diversity feature is used, the result set using a
greedy approach would be $\{A,B,C\}$. As may be seen, the latter ensures that the result objects are
not too similar among themselves.

uses Euclidean distance) could produce an output set $\{C,D,E\}$, a set of objects
that are very similar to each other. However, when the diversity feature is added, the
result set would be formed of C, B and A, which will be added to the result set in that
order. E and D would be disqualified due to them not satisfying the $divdist(.,.,.)$
constraint with the first object in the result set, C.

Diversity in Unordered Operators: Implementing diversity in unordered operators
is not so straightforward due to (a) absence of an ordering among objects in the result
set, and (b) since cardinality filters such as top-k are often not used on unordered
operators. Figure 4.7 illustrates the need for diversity in a skyline operator with the
query conveniently positioned at the origin for easy visualization; the usual skyline
operator would choose the result set $\{A,B,C,D\}$ of which three objects are very
similar to each other. Excluding any object for diversity consideration would distort
the semantics of the skyline result, and so would be the case if other objects need
to be brought in to the result set. Thus, the result set for the diversity enhanced
skyline operator is not well-defined. This is so in the case of some other unordered
operators too. However, one could use another orthogonal consideration that could
be used as a proxy for diversity and form a cardinality constrained result set. Such
a construction us used in the k most representative skyline operator [12]; here, a set
of k skyline points are identified such that the number of data objects dominated
by at least one of them is maximized. Though clearly different from the diversity
consideration, it may be highly correlated with diversity in an example such as that
in Figure 4.7 where $k = 2$ would yield a result set such as $\{A,D\}$. Nevertheless, the

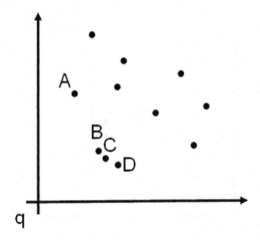

Fig. 4.7: Diversity Considerations in Skyline

point that we wanted to make is that diversity in unordered operators is not trivial, and operator specific considerations may be necessary many a time.

Motivation: Though not a popular consideration in similarity search systems yet, many readers may find that diversity is too intuitive to need motivation. Infact, diversity has been a major consideration in Information Retrieval literature (e.g., [1]). The idea is to reduce the redundancy in results so that the user finds the $n + 1^{th}$ result novel and interesting even after perusing the top-n results.

Examples: The most prominent work that has addressed diversity directly in similarity search is the KNDN operator [7] that modifies the top-k operator by incorporating the diversity feature. As outlined earlier, there have been operators that have used considerations similar to diversity as in the case of the k most representative skyline operator [12].

4.2.7 Summary of Features

Table 4.6 summarizes the various features that we have discussed so far in this section. All the features, with the exception of the indirection feature, need some extra information to work with, as illustrated in the Table. While we have attempted to cover most of the features that have been explored to enhance similarity operators (or makes sense to be used in combination with a variety of operators, even if they have not yet been used already), we should admit that this is in no way a comprehensive listing of the modifications that have been applied to operators. In fact, we have not looked at the variety of operator-specific features that have been explored such as k *most representative skyline* [12], *skyband* [15] and *geo-social skyline* [4].

Feature	Result Set and Semantics	Extra Info Required
Indirection ($\mathcal{O}^R$)	$\{x \mid q \in R_{\mathcal{O}}(x, \mathcal{X} \cup \{q\} - \{x\})\}$ - objects whose result set w.r.t $\mathcal{O}$ contains the query	None
Chromaticity ($\mathcal{O}^B$)	$R_{\mathcal{O}}(q, \{x \mid x.type \neq q.type\})$ - result of $\mathcal{O}$ restricted to work with a subset of objects that are not of the query type	Object types for each data object and query
Visibility ($\mathcal{O}^V$)	$R_{\mathcal{O}}(q, \{x \mid visible(x, q)\})$ - result of $\mathcal{O}$ restricted to work with a subset of object visible to the query	Set of obstacles to determine visibility
Multiple Queries ($\mathcal{O}^M$)	$s(.,.)$ vectors wrt multiple queries fused to form a single transformed representation $s_{\mathcal{O}M}$ followed by $\mathcal{O}$-style result determination	Multiple queries and a function to fuse multiple $s(.,x)$ vectors
Subspaces ($\mathcal{O}^S$)	$R_{\mathcal{O}}(q, \Pi_{subspace}(\mathcal{X}))$ - result of $\mathcal{O}$ restricted to work with the projection of $\mathcal{X}$ on attributes in $subspace$	Set of subspace attributes, $subspace$
Diversity ($\mathcal{O}^D$)	Result set formed so that a pairwise diversity constraint is respected	Specification of diversity attributes $divatts$, $mindiv$ - a minimum diversity threshold and $divdist$ - the diversity distance function

Table 4.6: Summary of Features

Nonetheless, we hope that this is useful as probably the first overview of general-purpose tools available to tune similarity operators for usage in various specialized scenarios.

4.3 Summary

In this chapter, we started with considering the classification of similarity operators based on two criteria. The first puts operators in one of two classes based on whether it produces *ordered* or *unordered* result sets, whereas the second considers the usage of attributes in the operator-specific similarity representations for an object. We looked at the implications of each of these choices and outlined operators that fall into each of these four classes. We then looked at features; a set of tools that are available for the designer of any search system to add to operators to tune the system to specific search needs. We outlined the semantics of the result set transformation under each of these features, illustrated motivating scenarios for the usage of such features, and listed operators from literature that have made use of them. As and when appropriate, we also commented on the practicality of the usage of such features, and implementation considerations in cases where the semantics of the feature-enhanced operator is not very clear. Through such a discussion of categorization of operators, and features that could be used along with operators, we hope that we have provided the interested reader with mental tools for quickly positioning operators with respect to the categories and the features they employ. Lastly, we hope that this discussion is also useful to the interested researcher in having outlined building blocks for similarity operators; these could now be combined in hitherto unexplored ways for usage in specialized scenarios, thus leading to advancing the frontier in the field of similarity search operators.

References

1. C. L. Clarke, M. Kolla, G. V. Cormack, O. Vechtomova, A. Ashkan, S. Büttcher, and I. MacKinnon. Novelty and diversity in information retrieval evaluation. In *Proceedings of the 31st annual international ACM SIGIR conference on Research and development in information retrieval*, pages 659–666. ACM, 2008.
2. E. Dellis and B. Seeger. Efficient computation of reverse skyline queries. In *Proceedings of the 33rd international conference on Very large data bases*, pages 291–302. VLDB Endowment, 2007.
3. P. M. Deshpande, P. Deepak, and K. Kummamuru. Efficient online top-k retrieval with arbitrary similarity measures. In *Proceedings of the 11th international conference on Extending database technology: Advances in database technology*, pages 356–367. ACM, 2008.
4. T. Emrich, M. Franzke, N. Mamoulis, M. Renz, and A. Züfle. Geo-social skyline queries. In *Database Systems for Advanced Applications*, pages 77–91. Springer, 2014.
5. R. Fagin, A. Lotem, and M. Naor. Optimal aggregation algorithms for middleware. *Journal of Computer and System Sciences*, 66(4):614–656, 2003.
6. Y. Gao, B. Zheng, G. Chen, W.-C. Lee, K. C. Lee, and Q. Li. Visible reverse k-nearest neighbor queries. In *Data Engineering, 2009. ICDE'09. IEEE 25th International Conference on*, pages 1203–1206. IEEE, 2009.
7. A. Jain, P. Sarda, and J. R. Haritsa. Providing diversity in k-nearest neighbor query results. In *Advances in Knowledge Discovery and Data Mining*, pages 404–413. Springer, 2004.
8. F. Korn and S. Muthukrishnan. Influence sets based on reverse nearest neighbor queries. In *ACM SIGMOD Record*, volume 29, pages 201–212. ACM, 2000.
9. Y. Kumar, R. Janardan, and P. Gupta. Efficient algorithms for reverse proximity query problems. In *Proceedings of the 16th ACM SIGSPATIAL international conference on Advances in geographic information systems*, page 39. ACM, 2008.
10. X. Lian and L. Chen. Monochromatic and bichromatic reverse skyline search over uncertain databases. In *Proceedings of the 2008 ACM SIGMOD international conference on Management of data*, pages 213–226. ACM, 2008.
11. X. Lian and L. Chen. Similarity search in arbitrary subspaces under l p-norm. In *Data Engineering, 2008. ICDE 2008. IEEE 24th International Conference on*, pages 317–326. IEEE, 2008.
12. X. Lin, Y. Yuan, Q. Zhang, and Y. Zhang. Selecting stars: The k most representative skyline operator. In *Data Engineering, 2007. ICDE 2007. IEEE 23rd International Conference on*, pages 86–95. IEEE, 2007.
13. Q. Liu, Y. Gao, G. Chen, Q. Li, and T. Jiang. On efficient reverse k-skyband query processing. In *Database Systems for Advanced Applications*, pages 544–559. Springer, 2012.
14. S. Nutanong, E. Tanin, and R. Zhang. Visible nearest neighbor queries. In *Advances in Databases: Concepts, Systems and Applications*, pages 876–883. Springer, 2007.
15. D. Papadias, Y. Tao, G. Fu, and B. Seeger. Progressive skyline computation in database systems. *ACM Transactions on Database Systems (TODS)*, 30(1):41–82, 2005.
16. M. Sharifzadeh and C. Shahabi. The spatial skyline queries. In *Proceedings of the 32nd international conference on Very large data bases*, pages 751–762. VLDB Endowment, 2006.

17. Y. Shi and B. Graham. A similarity search approach to solving the multi-query problems. In *Computer and Information Science (ICIS), 2012 IEEE/ACIS 11th International Conference on*, pages 237–242. IEEE, 2012.
18. Y. Tao, D. Papadias, and X. Lian. Reverse knn search in arbitrary dimensionality. In *Proceedings of the Thirtieth international conference on Very large data bases-Volume 30*, pages 744–755. VLDB Endowment, 2004.
19. Y. Tao, X. Xiao, and J. Pei. Efficient skyline and top-k retrieval in subspaces. *Knowledge and Data Engineering, IEEE Transactions on*, 19(8):1072–1088, 2007.
20. Q. T. Tran, D. Taniar, and M. Safar. Bichromatic reverse nearest-neighbor search in mobile systems. *Systems Journal, IEEE*, 4(2):230–242, 2010.
21. A. K. Tung, R. Zhang, N. Koudas, and B. C. Ooi. Similarity search: a matching based approach. In *Proceedings of the 32nd international conference on Very large data bases*, pages 631–642. VLDB Endowment, 2006.
22. A. Vlachou, C. Doulkeridis, Y. Kotidis, and K. Norvag. Monochromatic and bichromatic reverse top-k queries. *Knowledge and Data Engineering, IEEE Transactions on*, 23(8):1215–1229, 2011.
23. W. Wu, F. Yang, C.-Y. Chan, and K.-L. Tan. Finch: Evaluating reverse k-nearest-neighbor queries on location data. *Proceedings of the VLDB Endowment*, 1(1):1056–1067, 2008.
24. M. L. Yiu and N. Mamoulis. Reverse nearest neighbors search in ad hoc subspaces. *Knowledge and Data Engineering, IEEE Transactions on*, 19(3):412–426, 2007.
25. Z. Zhang, C. Jin, and Q. Kang. Reverse k-ranks query. *Proceedings of the VLDB Endowment*, 7(10), 2014.

Chapter 5
Advanced Operators for Similarity Search

We will now review a variety of advanced operators for similarity search. While we will consider a large number of operators that have been proposed in the last 10-15 years, the explosion of work that this area has seen during that timeframe means that we would have inevitably missed some work or the other however much we try to be comprehensive. Having said that, we cover most of the operators that have appeared in prominent venues. We will organize the discussion by categorizing the operators based on the basic similarity search operator that they build upon; for example, thus, an RkNN operator would fall under the weighted sum category, whereas dynamic skyline would fall under the skyline category. Weighted sum and Skyline happen to be the two basic operators that most others have been built upon. Accordingly, we will organize the discussion around these two categories, and discuss those that do not fall under these, in a separate section. Note that, though we describe the first category based on the weighted sum aggregation, in reality any scalar aggregate function such as Euclidean distance could be used instead.

5.1 Weighted Sum-based Operators

Before getting into the semantics of operators that build upon the weighted sum operator, we will briefly revisit the transformation involved in the weighted sum operator. Weighted sum transforms the similarity vector into a single scalar value that denotes the weighted sum of similarities across the attributes:

$$s_\mathcal{O}(q,x) = \sum_{1 \le i \le m} w_i \times s(q,x)[i] \tag{5.1}$$

where $[w_1, \ldots, w_m]$ is a vector of weights that is specified as part of the query, or pre-specified by the search system designer. The weighted sum function simply provides an ordering of objects with respect to the query; for example, the ordering of objects based on the non-ascending order of $s_\mathcal{O}(q,.)$ provides a proximity-based ordering that puts the objects most similar to the query at the top of the ordering.

© The Author(s) 2015
D.P and P.M. Deshpande, *Operators for Similarity Search*,
SpringerBriefs in Computer Science, DOI 10.1007/978-3-319-21257-9_5

Since not all objects in the dataset can be displayed in any practical system, a filter function is typically used on top of this. We now revisit the top-k and range filters that are often used with the weighted sum operators, by outlining the result set with respect to them:

$$R_{\mathscr{O}}(q, \mathscr{X}) = \begin{cases} argmax_{X \subseteq \mathscr{X} \wedge |X|=k} \ \sum_{x \in X} s_{\mathscr{O}}(q,x) & if \ \mathscr{O} = \text{weighted sum top-k} \\ \{x | x \in \mathscr{X} \wedge s_{\mathscr{O}}(q,x) \leq \tau\} & if \ \mathscr{O} = \text{weighted sum range} \end{cases}$$

(5.2)

Informally, the top-k operator chooses the k elements with the highest $s_{\mathscr{O}}(q,x)$ values, and the range operator chooses all objects that have $s_{\mathscr{O}}(q,x)$ evaluating to less than a threshold τ.

While describing the various operators that build upon the weighted sum formulation in this section, we will use a tabular representation to illustrate (a) whether the operator uses ordering in the result set, (b) the nature of the filter used (e.g., top-k, range etc.) and (c) the features (from the previous chapter) that the operator uses, if any. An example twin-table representation for the weighted sum top-k operator appears in Table 5.1. The operator would have a tick mark in the appropriate combination of (a) and (b) in the left table, and tick marks against all the features employed in the right table. In certain cases where a rare feature that has not yet been introduced gets used, we will indicate that as an additional row in the feature table. We will now proceed to introducing the various weighted sum-based operators in respective subsections hereon. As a short-hand to represent results, we will use $\mathscr{O}(q, \mathscr{X})$ to refer to the result set of the operator $\mathscr{O}$ with query q on dataset $\mathscr{X}$; i.e., $\mathscr{O}(q, \mathscr{X}) = R_{\mathscr{O}}(q, \mathscr{X})$; thus $top\text{-}k(q, \mathscr{X})$ refers to the result of the (weighted sum) $top\text{-}k$ operator.

Table 5.1: Weighted Sum Top-k Operator

Base Operator	Ordered Results	Unordered Results
Top-k Filter	✓	
Range Filter		
Other Filter		

Indirection	✗	Bichromaticity	✗
Visibility	✗	Multiple Queries	✗
Subspaces	✗	Diversity	✗

Table 5.2: Reverse k Nearest Neighbor

Base Operator	Ordered Results	Unordered Results
Top-k Filter	✓	
Range Filter		
Other Filter		

Indirection	✓	Bichromaticity	✗
Visibility	✗	Multiple Queries	✗
Subspaces	✗	Diversity	✗

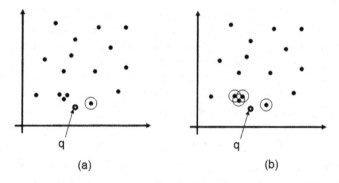

Fig. 5.1: RkNN Example. (a) Dataset showing result objects (circled) for a query with $k = 1$. (b) Same dataset and query, with results for $k = 3$ indicated.

5.1.1 Reverse k Nearest Neighbor (RkNN): Indirection on Top-k

This operator is formed by using the indirection feature on the results of the top-k operator; the name for the operator is derived from the alternative name for the top-k operator, i.e., the k nearest neighbor operator. The specification of the result set for RkNN is the following:

$$RkNN(q, \mathcal{X}) = \{x | x \in \mathcal{X} \land q \in \text{top-k}(x, \mathcal{X} \cup \{q\} - \{x\})\} \qquad (5.3)$$

Thus, all objects that have q among their top-k results, when run on a dataset where the query is interchanged with the object, form part of the result set for the RkNN operator. The tabular representation for this operator appears in Table 5.2. It may be noted that the left table illustrates the basic operator that the RkNN operator builds upon. While the basic operator produces an ordered result set with a top-k cut-off, the results of the RkNN operator itself is unordered, and uses a crisp binary membership function that does not need any cut-off. This is so due to the construction of the indirection feature as we will illustrate by an example. The special case of the RkNN query where k is set to 1, called the RNN query, was introduced in [11]. The general version was also studied soon after, in [22].

Example: Figure 5.1 shows a small dataset on two dimensions with point objects where the similarity between any pair of objects is inversely related to the distance between the objects in the figure. In Figure 5.1(a), the RkNN operator with $k = 1$ (i.e., RNN) is shown to produce a single object in the result set. It is interesting to note the difference in the semantics of the top-1 operator and the RNN operator by way of observing that the cluster of three objects, despite being closer to the query than the result object, are not part of the result. This is so since the cluster of three objects are so close together that for each of them, the nearest neighbor would be another object in the cluster. The presence of an object nearer to each of them than the query leads to excluding them from the RNN result. The scenario changes when we set $k = 3$ as may be seen in Figure 5.1(b). Since the cluster is composed of only three objects, for every object, the cluster itself can contribute only two proximal neighbors. In particular, the third neighbor would need to be outside the cluster; in this figure, it so happens that the query is the third nearest neighbor for all objects in the cluster. Thus, the RkNN result set for $k = 3$ is a set of four objects, the cluster and the object that was in the RNN result set. It may be noted in this context that, that the RkNN result set grows with increasing k; i.e.,

$$RkNN_{k=k_1}(q, \mathcal{X}) \subseteq RkNN_{k=k_2,k_2 \geq k_1}(q, \mathcal{X}) \tag{5.4}$$

Motivation: Consider the case of resource planning for resiliency where each resource, when unavailable, may be replaced by one of its top-k most similar resources. Then, the RkNN set for a resource determines the set of objects for which it could function as a back-up. Thus, in scenarios such as de-commissioning a resource, one needs to be aware of the size of the RkNN set of that resource; this is so since each resource in the RkNN set would be impacted as its list of backup resource would now have to be updated with a less similar resource.

Table 5.3: Bichromatic Reverse k Nearest Neighbor

Base Operator	Ordered Results	Unordered Results
Top-k Filter	✓	
Range Filter		
Other Filter		

Indirection ✓	Bichromaticity	✓
Visibility ✗	Multiple Queries	✗
Subspaces ✗	Diversity	✗

5.1.2 Bichromatic Reverse k Nearest Neighbor (BRkNN)

As the name indicates, this operator is nothing but the RkNN operator enhanced with the feature of bichromaticity, as illustrated in the tabular representation in Table 5.3. The bichromaticity feature assumes that data objects in $\mathcal{X}$ as well as the query be associated with one of two types; in many common scenarios, these types are *service providers* and *consumers*. The result set may be formally specified as the following:

$$BRkNN(q, \mathcal{X}) = \{x | x \in \mathcal{X}_{type \neq q.type} \wedge q \in top\text{-}k(x, \mathcal{X}_{type=q.type} \cup \{q\})\} \tag{5.5}$$

where $\mathcal{X}_{type=t}$ and $\mathcal{X}_{type \neq t}$ denote the subset of objects in $\mathcal{X}$ that belong to the type t or not respectively. Procedurally, the BRkNN set may be derived by issuing a top-k operation using each object in $\mathcal{X}_{type \neq q.type}$ as the query, where the top-k operation is performed on the subset of data objects that belong to the query type, as well as the BRkNN query object. If the BRkNN query object itself belongs to the top-k result set, the corresponding data object is added to the BRkNN result set. This naive and obviously inefficient procedural approach just serves to present the

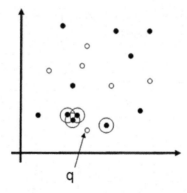

Fig. 5.2: BRkNN Example. The two types are indicated by differently shaded data points. The
results of the BRkNN query with $k = 1$ are circled.

semantics of the *BRkNN* operator; actual implementations could significantly differ
while adhering to the semantics of the naive approach above. The *BRkNN* query was
first considered in [11] for the special case of $k = 1$; it has since attracted plentiful
attention.

Example: Figure 5.2 shows a variant of the earlier example, where data objects are
shaded differently according to their types. Consider the *BRkNN* query with $k = 1$
on the white query object indicated. The results of the *BRkNN* query are those black
objects that do not have a white neighbor nearer to it than the query. Thus, unlike the
basic (i.e., monochromatic) RkNN case we saw earlier, the black objects in the three
object cluster do not get counted as neighbors to each other, since the top-k query
on them is directed towards only the white objects. Accordingly, the four circled
objects are returned as results.

Motivation: The *BRkNN* query comes in handy when the database contains service
providers as well as consumers. It is of interest for a service provider to find the
consumers whose interest profiles are such that the service provider itself is among
the best matching service providers for them. If the profiles are purely geographic,
it amounts to finding the consumers for whom the service provider is among the
top-k closest service providers. The service provider could target such customers
to send offers to, since it is likely that they would respond positively, due to the
service provider being in their proximity. The *BRkNN* set also may be called the
influence set, since those are the consumers who can be easily influenced by the
service provider.

Table 5.4: Reverse k Farthest Neighbor

Base Operator	Ordered Results	Unordered Results
Top-k Filter		
Range Filter		
Other Filter	Bottom-k Filter	

Indirection	✓	Bichromaticity	✗
Visibility	✗	Multiple Queries	✗
Subspaces	✗	Diversity	✗

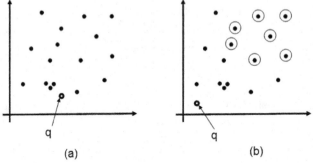

(a)   (b)

Fig. 5.3: RkFN Example with $k = 1$. (a) and (b) show different queries and their RkFN result sets. For the first query, there are no objects in the RkFN result set, whereas the second query returns a RkFN result set with many elements.

5.1.3 Reverse k Farthest Neighbor (RkFN): Farthest instead of Nearest

The RkFN operator differs from the RkNN operator only in the filter that is employed; instead of choosing the top-k similar objects to the query in the basic operator, the RkFN operators builds upon a filter that chooses the k objects that are least similar to the query. We call this filter as the bottom-k filter, as indicated in Table 5.4. We will now introduce the formal definition for the weighted sum bottom-k operator:

$$WSBk(q, \mathscr{X}) = \underset{X \subseteq \mathscr{X} \wedge |X| = k}{argmin} \sum_{x \in X} s_{\mathcal{O}}(q, x) \qquad (5.6)$$

where $WSBk(q, \mathscr{X})$ denotes the result set for the weighted sum bottom-k as the set of k least similar objects to the query. Having defined the bottom-k filter, the $RkFN$ result set is the following:

$$RkFN(q, \mathscr{X}) = \{x | x \in \mathscr{X} \wedge q \in WSBk(x, \mathscr{X} \cup \{q\} - \{x\})\} \qquad (5.7)$$

Thus, analogous to the $RkNN$ operator, the $RkFN$ operator chooses those objects that have the query among their respective bottom-k result set by usage of the indirection feature. The $RkFN$ query, for the special case of $k = 1$, was first considered

in [11]. Since then, the full *RkFN* query has been considered in scenarios such as
GIS applications [12].

Example: Figure 5.3 illustrates two different query points and their *RkFN* result
sets for $k = 1$ on the same example dataset. For the first case, there are no objects
that have the query as the farthest object. In general, this would be true for queries
that are not at the *corners* of the dataset (when the dataset may be projected on a
Euclidean space as in the figure); for example, when the query object is reasonably
central in the dataset, any data object would be able to find a farther point by extending the search from the query in the direction opposite to the object in question.
Figure 5.3(b) shows a query whose *RkFN* result set has many objects.

Motivation: In the case of organizational resiliency planning, a list of similar resources would be maintained for every resource. In cases where a resource needs
to be taken down for maintenance, one of the similar resources could be used as a
backup. In many cases, these lists could be hand-curated to adjust for various constraints; for example, a server running *Windows* may not be chosen as a candidate
backup server for a *Linux* machine even if there is high similarity on other attributes.
The *RkFN* result for a particular resource is intuitively a set of resources for whom
the resource in question is very unlikely to be in the backup list. Thus, the size of
the *RkFN* list is inversely related to the number of resources it could impact. The
bichromatic variant of *RkFN* is also useful since the *RkFN* set for a service provider
lists the set of consumers it is least likely to influence.

Table 5.5: Constrained Nearest Neighbor

Base Operator	Ordered Results	Unordered Results			
Top-k Filter	✓		Indirection	✗	Bichromaticity	✗
Range Filter			Visibility	✗	Multiple Queries	✗
Other Filter			Subspaces	✗	Diversity	✗
			Constraints on Results			✓

5.1.4 Constrained Nearest Neighbors: Results must satisfy Constraints

The constrained nearest neighbor operator is a simple modification over the top-k
operator, where a constraint is expected to be satisfied for each object in the result
set as indicated in Table 5.5. The declarative specification of the result set is as
follows:

$$CNN(q, \mathcal{X}) = \underset{x \in \mathcal{X} \wedge Satisfies(x,C)}{argmax} s_{\mathcal{O}}(q,x) \tag{5.8}$$

where $Satisfies(x,C)$ is a boolean function that determines whether the data object x satisfies the constraint set C. The set of constraints, C, is extrinsic to the dataset

and could be specified by the user or pre-specified in the search system. The *CNN* operator may be easily generalized to the *CkNN* operator by allowing to choose the top-k similar objects in $\mathcal{X}$ that satisfies the constraints.

$$CkNN(q, \mathcal{X}) = \underset{X \subseteq \mathcal{X} \wedge Satisfies(X,C) \wedge |X|=k}{argmax} \sum_{x \in X} s_{\mathcal{O}}(q,x) \qquad (5.9)$$

Here, the *Satisfies*(.,.) function is overloaded with a set operand, and the semantics generalized to checking whether *each* object in the set satisfy the constraints in *C*. The constraints could be of any form since the above definition only requires that the check for whether a data object satisfies the constraint be well-defined; some examples of types of constraints are as follows:

- **Value Constraints:** This constraint enforces that the object in question take certain specific values (or ranges of values) for particular attributes. For example, using value constraints, the search can be directed to people entities who are within a particular age range, or reside within a particular country.
- **Inclusion within a Polygon:** In data that can be embedded within a Euclidean space or in the case of geographic data that can be visualized in 2-d, a constraint could be used to check whether the data object is within a polygonal area that is specified as part of the constraints. Such a constraint could be used to check for hotels within a particular state, where the geographic region of the state may be approximated using a polygon.

Constrained nearest neighbor queries were explored in [6], where two approaches to process them were considered; the first is to process the regular nearest neighbor query in incremental fashion and checking for constraint satisfaction before emitting the results, whereas the second restricts the search for nearest neighbors to the data subset that satisfies the constraint. The choice between the two approaches depends upon factors such as the selectivity of the constraint.

Example: Figure 5.4 shows two datasets for constrained nearest neighbor, with two different constraints. In Figure 5.4(a), the area of the dataset satisfying the constraint is the entire quadrant to the north-east of the query object, wheres Figure 5.4(b) uses a smaller rectangular region as the constraint area.

Motivation: In the case of queries from a moving car to find nearby restaurants, it might make sense to constrain the search results to the direction of the movement. For example, while driving southward, the user may be interested in finding restaurants that are south of the current location since even a very closeby restaurant that is north would require the driver to change direction and drive in the opposite direction to the final destination. Constraints on age are useful while searching for suitable candidates for a job posting that has an age constraint.

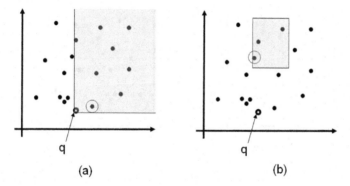

Fig. 5.4: CNN Example. (a) shows a constraint set where the search is directed to the north-east of the query entity, whereas the constraint in (b) checks for inclusion in a rectangular area. In both cases, the dataset sub-area satisfying the constraint is shaded in grey.

Table 5.6: Visible kNN

Base Operator	Ordered Results	Unordered Results
Top-k Filter	✓	
Range Filter		
Other Filter		

Indirection	✗	Bichromaticity	✗
Visibility	✓	Multiple Queries	✗
Subspaces	✗	Diversity	✗

Table 5.7: Visible RkNN

Base Operator	Ordered Results	Unordered Results
Top-k Filter	✓	
Range Filter		
Other Filter		

Indirection	✓	Bichromaticity	✗
Visibility	✓	Multiple Queries	✗
Subspaces	✗	Diversity	✗

5.1.5 Visible kNN and RkNN: Visibility Constraints on Top-k and RkNN

We will now consider two operators, $VkNN$ and $VRkNN$, those that employ the visibility feature on the top-k and $RkNN$ operators respectively. Visibility is a special set of constraints that is useful in scenarios such as geographic data. The result set for the $VkNN$ operator is defined as follows:

$$VkNN(q, \mathcal{X}) = \underset{X \subseteq \mathcal{X} \wedge |X| = k \wedge (\forall x \in X, visible(q,x))}{argmax} \sum_{x \in X} s_{\mathcal{O}}(q,x) \qquad (5.10)$$

where $visible(q,x)$ evaluates to true if the straight line connecting q and x is not intercepted by an obstacle in a set of obstacles $Obst$ (as seen in the earlier chapter). Thus, the $VkNN$ operator chooses the top-k similar objects that are visible to the query. The tabular representation for the $VkNN$ operator appears in Table 5.6. The $VRkNN$ operator extends $VkNN$ by using the indirection feature:

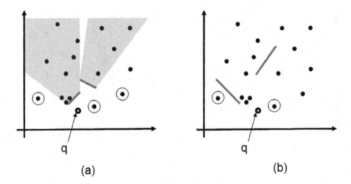

Fig. 5.5: (a) *VkNN* Example ($k = 3$). The obstacles are shown as grey thick lines, and the area that each obstacle shields from the query is shown in light grey. The three closest neighbors that are not intercepted by the obstacles are chosen as the results, as indicated by the circle around them. (b) *VRkNN* Example ($k = 1$). This shows a different set of obstacles and the result set for the *VRkNN* query wrt them. The closest neighbors for the leftmost object are the three object cluster, which in this case, are not visible from the object due to the obstacle. Thus, the query point forms the nearest visible neighbor, bringing the leftmost point into the result set.

$$VRkNN(q, \mathscr{X}) = \{x|x \in \mathscr{X} \wedge q \in VkNN(x, \mathscr{X} \cup \{q\} - \{x\})\} \qquad (5.11)$$

Informally, *VRkNN* chooses the set of objects that have the query in their respective *VkNN* result sets (Refer Table 5.7). Initial proposals of the *VkNN* and *VRkNN* operators appear in [17] and [8] respectively. It may be noted that visibility constraints by obstacles can be transformed into equivalent polygonal inclusion constraints for processing by the constrained nearest neighbor queries; however, in the presence of a large number of obstacles, the polygons become very complex.

Example: Figure 5.5(a) illustrates an example of a dataset with two obstacles. The shaded region shows the area that is shielded from the query by way of each of the obstacles, and the resulting *VkNN* result set for $k = 3$ is indicated by circles around the result objects. Figure 5.5(b), on the other hand, illustrates the results for the *VRkNN* query with $k = 1$ with respect to a different set of obstacles.

Motivation: Obstacles largely make sense only in cases where the objects are represented in a geographical space. Visibility of a restaurant from a house is useful from the restaurant's perspective since it could help in increasing the chances that the resident would visit it. Visibility is critical in the case of positioning billboards since large obstacles could reduce the number of people who can view the billboard, thus affecting the impact of the advertisement being placed on it.

Obstacle Nearest Neighbors: Apart from using visibility constraints induced by obstacles, there has been some work on nearest neighbor queries where the distance between two objects is estimated in such a way that obstacles are accounted for. For example, if there is a wall in the path between two objects, the *obstacle distance* between them would be the shortest distance through other paths including those that just wrap around the wall. The obstacle nearest neighbor query [26] identifies objects that are at the shortest obstacle distance from the query. It may be noted that obstacle distance can be considered as a another type of aggregation function to fit it in the weighted sum framework.

Table 5.8: Subspace Top-k

Base Operator	Ordered Results	Unordered Results
Top-k Filter	✓	
Range Filter		
Other Filter		

Indirection	✗	Bichromaticity	✗
Visibility	✗	Multiple Queries	✗
Subspaces	✓	Diversity	✗

Table 5.9: Subspace Range

Base Operator	Ordered Results	Unordered Results
Top-k Filter		
Range Filter		✓
Other Filter		

Indirection	✗	Bichromaticity	✗
Visibility	✗	Multiple Queries	✗
Subspaces	✓	Diversity	✗

5.1.6 Subspace Top-k and Range Queries

The subspace top-k and range operators are modifications of the respective basic operators by usage of the subspaces feature. The subspace feature entails that the similarity search query be performed on a subset of attributes in the schema, where the attributes in the subspace may be specified at query time. The semantics of these operators, as specified in Tables 5.8 and 5.9 respectively, are formally outlined as follows:

$$SubspaceTop\text{-}k(q, \mathscr{X}) = Top\text{-}k(q, \Pi_{subspace}(\mathscr{X})) \qquad (5.12)$$

$$SubspaceRange(q, \mathscr{X}) = range(q, \Pi_{subspace}(\mathscr{X})) \qquad (5.13)$$

where $\Pi_{subspace}(\mathscr{X})$ denotes the dataset projected on to the attributes in the set of attributes denoted by *subspace*. The query q is also accordingly specified only on the attributes in *subspace*. The Subspace Top-k and Range operators are simply the respective operators when instantiated on the specific subspaces specified along with the query. [23] and [14] describe approaches to handle the subspace top-k and range operators respectively. It may be noted that, under the weighted sum formulation, choosing a subset of attributes is equivalent to setting the weights of other

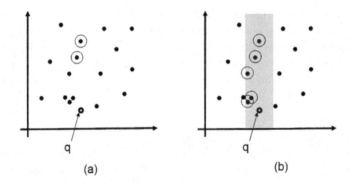

Fig. 5.6: (a) shows the results of a subspace top-k query with $k = 2$ on a subspace comprising of only the X attribute. (b) highlights the region searched over as well as the results of a subspace range query where the only attribute selected is X.

attributes to 0.0 in the weight vector used; however, these pose several new index-ing challenges since indexes built across all attributes are not optimized for queries where only a subset of attributes need to be considered.

Example: Figure 5.6 illustrates the results of subspace queries where only the X attribute is selected for inclusion in the subspace. Figure 5.6(a) illustrates the two objects that are closest to the query on the X attribute, which consequently form the result set of the subspace top-k query. Figure 5.6(b) illustrates that objects are chosen solely based on their X-attribute proximity to the query regardless of their Y-attributes, in the corresponding subspace range query.

Motivation: The argument for allowing for differential weighting of attributes by means of a weight vector is that different scenarios pose different prioritization of attributes. Subspaces is just a means of selecting an extreme value, specifically a value of 0.0, for attributes not in the chosen subspace. A person searching for hotels to stay in during a business trip may choose to rely solely on attributes such as proximity to the airport and availability of a restaurant, whereas price and proximity to the beach may be the considerations for a vacation planning context. Such queries may be accomplished by choosing the respective sets of attributes as the chosen subspace, in a subspace similarity search operator.

Table 5.10: K Nearest Diverse Neighbors

Base Operator	Ordered Results	Unordered Results
Top-k Filter	✓	
Range Filter		
Other Filter		

Indirection	✗	Bichromaticity	✗
Visibility	✗	Multiple Queries	✗
Subspaces	✗	Diversity	✓

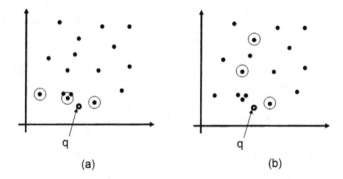

Fig. 5.7: (a) shows the results of KNDN ($k = 3$) when both the attributes are chosen within *divatts*. As may be seen, only one object from the three object cluster may be chosen for inclusion in results due to the diversity constraint. (b) indicates the results for the same operator when *divatts* $= \{Y\}$. Here, the results are seen to be have more spread on the Y attribute.

5.1.7 K Nearest Diverse Neighbors: Diversity-conscious Top-k

The *K Nearest Diverse Neighbors* (KNDN) operator also uses a constraint that the results should satisfy; however, unlike the *CkNN* operator, this constraint is on the set rather than on individual objects. The result set may be specified as:

$$KNDN(q,\mathscr{X}) = \underset{X\subseteq\mathscr{X}\wedge|X|=k\wedge(\forall x_1,x_2\in X,x_1\neq x_2,divdist(x_1,x_2,divattrs)\geq mindiv)}{argmax} \sum_{x\in X}s_\mathscr{O}(q,x)$$
$$(5.14)$$

The third predicate in the condition is the new one introduced by the diversity feature. Informally, it enforces that for every pair of distinct objects within the result set, the distance between them on a chosen subset of attributes, *divatts*, needs to be above a threshold, *mindiv*. The *KNDN* operator simply chooses the set of k objects that together have the highest aggregate similarity to the query, while enforcing the diversity constraint. Sicne identifying the optimal result set is typically hard given the nature of the objective function, greedy approaches are often used. As outlined in the previous chapter when the diversity feature was introduced, $divdist(.,.)$ may use any simple aggregation such as weighted sum aggregation of distances over the attributes in *divatts*. However, the paper proposing this operator chose to use an aggregation that uses a high weighting for attributes on which the specific pair of objects under consideration has low similarities; we will refer the interested reader to the details in [9].

Example: Figure 5.7(a) shows the results of the KNDN operator ($k = 3$) for two choices of *divatts*. With *divatts* $= \{X,Y\}$, only one of the objects in the three object cluster can be chosen among the results since inclusion of a second object from the cluster would create a pair within the result set that violates the diversity constraint.

Figure 5.7(b) shows that when $divatts = \{Y\}$, none of the objects from the cluster would form part of the result set since they are too close to the closest neighbor on the Y attribute. Due to the diversity constraint, the search for KNDN results extends further away from the query point on the Y attribute.

Motivation: Diversity in search reduces monotony among the top results. In the case of a scenario where one is searching for a product that is similar to the query product, it would be less informative to show products that are very similar to each other compared to showing products that differ substantially from each other. This is so since the latter set of results are likely to enable the user get a feel for the different kinds of variability that exist in the space. Infact, diversity has been a well-studied issue in information retrieval [2] for such reasons.

Table 5.11: Multi-Query Top-k

Base Operator	Ordered Results	Unordered Results
Top-k Filter	✓	
Range Filter		
Other Filter		

Indirection	✗	Bichromaticity	✗
Visibility	✗	Multiple Queries	✓
Subspaces	✗	Diversity	✗

5.1.8 Multi-Query Top-k

The multi-query top-k operator allows the user to specify multiple queries. Each data object is ranked with respect to the similarity it has with respect the multiple queries in the set. This builds upon the regular top-k operator as shown in Table 5.11. As we saw in the earlier chapter, a simple method to specify a multi-query similarity search operator would be to specify the aggregate similarity of a data object to multiple queries by way of simple aggregation functions such as a sum or average of similarities across queries. However, the multi-query top-k operator proposed in [21] uses a quite complex aggregation of similarities across queries. In the interest of looking at operators that have been tested empirically, we will focus on the construction of the multi-query top-k result set in [21].

Consider a set of queries specified as $Q = \{q_1, q_2, \ldots\}$; within this, each query is much like a query object that we have seen in the single query scenario. Further, a weight may be specified for each query, with W_l denoting the weight for the lth query. Now, for every attribute a_j in $\mathscr{A}$ and for every query q_l in Q, we will order the data objects in the non-decreasing order of distance with respect to that attribute. Let PS_{lj} be the top-p data objects - i.e., the p most similar data objects to the query q_l on attribute a_j.

$$PS_j = \bigcup_l PS_{lj} \tag{5.15}$$

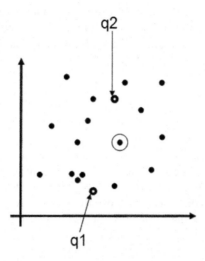

Fig. 5.8: The Multi-Query Top-k Operator with two queries showing result set with $k = 1$.

Thus, PS_j has the set of data points from $\mathscr{X}$ that are among the top-p nearest neighbors to *at least* one of the queries in Q, on the attribute a_j. We are now in a position to outline the multi-query distance for each data point $x \in \mathscr{X}$.

$$MQD(Q,x) = \frac{\sum_{a_j \in \mathscr{A}} \left(d(\hat{Q},x)[j] \times I(x \in PS_j) \right)}{\left(\sum_{a_j \in \mathscr{A}} I(x \in PS_j) \right)^2} \tag{5.16}$$

where $\hat{Q}$ is the weighted average of the queries in Q, weighted according to the weights W_l. $I(.)$ is the indicator function that is set to 1 and 0 if the boolean expression within it evaluates to true and false respectively. Thus, the MQD is the aggregate of distances on those attributes on which x is among the top-p similar objects to any one of the query. If x belongs to the top-p lists of many queries, the MQD is discounted by way of a large denominator. It may be noted that if p is set to the size of the dataset, MQD reduces simply to the ranking of objects based on the distance to $\hat{Q}$. The k objects with the smallest MQDs are then output as the results:

$$MultiQueryTop\text{-}k(Q, \mathscr{X}) = \underset{X \subseteq \mathscr{X} \wedge |X| = k}{argmin} \sum_{x \in X} MQD(Q,x) \tag{5.17}$$

Example: Figure 5.8 shows the result of the multi-query top-k operator with two queries for $k = 1$. With equal weights between queries, a data object that is close to the mid-point of the line connecting the two queries would be chosen as the nearest neighbor, as indicated in the figure.

Motivation: Multiple queries are common in many real world scenarios; consider the example of a person who is trying to join a gym and wants to find one that is equidistant from both home and office since she may have to go to to the gym from either of these locations. The home and office form multiple query points in a geographic space, and the multi-query top-k operator would then prioritize gyms that are equidistant from these two locations.

5.1.9 Summary

This brings us to the end of our selection of advanced operators that build upon the weighted sum framework. As we have seen, all the features discussed in the previous chapter have been used on weighted sum operators. The larger body of work has been on extending the top-k operator as compared to the range operator; this is probably so since the top-k operator allows the user to control the size of the output set, making it applicable across a wide spectrum of scenarios ranging from displaying results on a mobile device to identifying a large set of customers for targeted marketing.

5.2 Skyline-based Operators

We will now consider the operators that build upon the basic skyline operator that was introduced in an earlier chapter. Let us briefly revisit the skyline result set semantics before going into operators based on it.

$$Skyline(q, \mathcal{X}) = \{x | x \in \mathcal{X} \wedge (\nexists y \in \mathcal{X}, y \succ_q x)\} \tag{5.18}$$

Thus, the skyline result set comprises all objects in $\mathcal{X}$ such that there is no object y dominating them with respect to the query, the domination relationship with respect to q denoted by $\succ_q$. The domination relationship is based on the distance vectors $d_\mathcal{O}(q,x)$:

$$y \succ_q x = \begin{cases} true, & (\forall a_j \in \mathcal{A}, d_\mathcal{O}(q,y)[j] \leq d_\mathcal{O}(q,x)[j]) \wedge \\ & (\exists a_j \in \mathcal{A}, d_\mathcal{O}(q,y)[j] < d_\mathcal{O}(q,x)[j]) \\ false, & otherwise \end{cases} \tag{5.19}$$

Thus, y is said to dominate x with respect to q if y is (a) at most as distant from q as x on *each* element of the $d_\mathcal{O}(.,.)$ vector, and (b) is strictly less distant than x from q on at least one element of the $d_\mathcal{O}(.,.)$ vector. The second condition helps to ensure that duplicate objects (i.e., those that have the same value for each attribute) do not dominate each other. These two conditions are represented as separate conditions on the first case in the equation above. It may be recalled that the typical $d_\mathcal{O}(.,.)$ vector

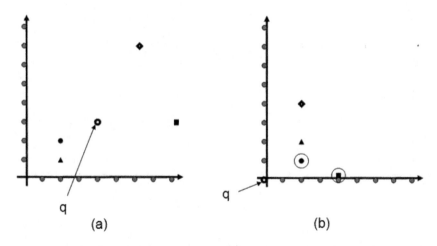

Fig. 5.9: (a) shows a small dataset of four data objects along with the query object. (b) plots the $d_\mathcal{O}(.,.)$ vectors of the respective data points on the same 2-d space, and indicates the results of the skyline query by circles.

used in skyline processing is precisely the $d(.,.)$ vector; given that equivalence, each element of the $d_\mathcal{O}(.,.)$ vector maps to one attribute of the schema $\mathscr{A}$.

Research on inventing operators based on the skyline operator has considered very many possibilities. This makes it hard to organize each advanced skyline operator in the form of a table as we did for the weighted sum-based operators. Hence, while we will stick to the semantics-example-motivation narrative in introducing skyline operators too, we will not use the tabular form representations. Many of the advanced skyline operators target the query-less scenario, where the attributes are all ordinal (e.g., numeric), and the operator is applied with an implicit query that takes an extreme value on each attribute; for example, the implicit query for a hotel search would be $Rating = 5 * \wedge Cost = 0.0$. The semantics of all such operators can be adapted to be used in query-based scenarios. Since we are interested in similarity search which involves comparing objects against a query, we will discuss the query-based adaptations of such operators in the following subsections.

Example: We now illustrate the skyline operation with an example in Figure 5.9; we will use a small dataset to ensure easy understanding. Consider the dataset of four data objects in the 2-d euclidean space in Figure 5.9(a). The data points are shown using different types of symbols such as *triangle, diamond, square* and *circle*. We have included a grey point on the axes after every unit of distance to indicate the exact positioning of each data point. Thus, the data points take on the following co-ordinates: $(2,1)$ (triangle), $(2,2)$ (circle), $(6,7)$ (diamond) and $(8,3)$ (square). The query point is at $(4,3)$. We will use the absolute numeric difference between attribute values to compare objects. Thus, the $d_\mathcal{O}(.,.)$ vector for the triangle object would be $[2,2]$ whereas that for the diamond object would be $[2,4]$. Such distance

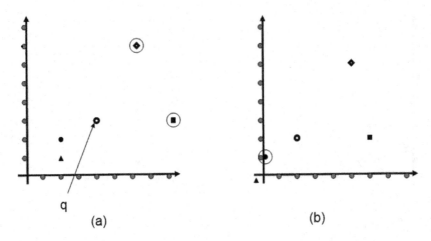

Fig. 5.10: (a) shows the reverse skyline result set for the query. (b) shows the plot of the $d_)(.,.)$ vector as well as the skyline results when the triangle object is used as a query. It may be seen that the circular object dominates the query; this causes the query object to be excluded from the skyline result set for the triangle object, and consequently causes the triangle object to be omitted from the reverse skyline result set for the query.

vectors are plotted in Figure 5.9(b) with the query now being re-positioned at the origin since the distance vectors are computed with respect to it. The result of the skyline query would now be the objects denoted by *circle* and *square*. This is so since the circle dominates the triangle as well as the diamond; though it is also clear that the triangle also dominates the diamond, it is enough to have one dominator for an object to be pushed out of the skyline result set. We will use variants of this dataset to illustrate the advanced skyline operators in this section, wherever appropriate.

5.2.1 Reverse Skyline: Indirection on the Skyline

The reverse skyline operator uses the indirection feature on the basic skyline operator. The result set is thus specified as the following:

$$ReverseSkyline(q, \mathscr{X}) = \{x | x \in \mathscr{X} \land q \in Skyline(x, \mathscr{X} \cup \{q\} - \{x\})\} \quad (5.20)$$

Similar to the usage of indirection seen earlier, the Reverse Skyline result set comprises all objects where the query is part of their respective skyline results. Reverse skyline query processing has been considered in various scenarios [3, 4].

Example: Figure 5.10(a) illustrates the reverse skyline result set for the query indicated. It may be noted that only the diamond and square object form part of the reverse skyline result set. This is because when either of the triangle or the circle object are considered, the other dominates the query with respect to them. An example appears in Figure 5.10(b) where the $d_\mathcal{O}(.,.)$ vectors are plotted when the triangle object is used as the query. It is clear in that example that the circular object would dominate the query when the domination relationship is assessed with respect to the triangle object.

Motivation: We will once again use a resiliency planning scenario to motivate the reverse skyline query; consider the case of assigning backup lists for resources where the weighting between the various attributes is not known beforehand. The skyline result set will necessarily include the most similar resource for any weighting of attributes. Thus, the skyline set would form a reasonable list to choose backups from, when a resource needs to be taken off. The reverse skyline set of a resource then denotes the resources for which it could function as a backup. Those resources with large reverse skyline sets are then critical since they are among the preferred backups for a large number of resources. Such critical servers may be maintained with a higher degree of attention or caution.

5.2.2 Thick Skyline: Skyline Objects and their Neighbors

The notion of thick skylines, introduced in [10], generalizes the skyline result set by inclusion of objects that are close to skyline objects. The Thick Skyline result set may be formally defined as:

$$TS(q, \mathcal{X}) = Skyline(q, \mathcal{X}) \cup \{x | x \in \mathcal{X} \wedge (\exists y \in Skyline(q, \mathcal{X}), x \in WSR_\varepsilon(y, \mathcal{X}))\}$$
$$(5.21)$$

The thick skyline comprises all the skyline objects, and additionally, includes objects that are within an ε distance away from an object in the skyline. In the above equation, WSR_ε indicates the weighted sum range query results where the range is chosen to be ε.

Example: Figure 5.11 illustrates the thick skyline result set with respect to the query. The triangle object is included due to being within ε distance of the circle object (which is in the simple skyline) by way of the second condition in the $TS(.,.)$ equation above.

Motivation: Consider a hotel search scenario where one would like to choose among the skyline hotels in two attributes *star rating* and *price*. The implicit query could be the ideal hotel with *five star* rating and *zero* cost. It is possible, due to extrinsic reasons such as a conference happening at one of the skyline hotels, that the chosen hotels are not able to accept reservations. The thick skyline includes some additional options similar to the skyline hotel, for usage in such scenarios.

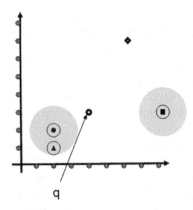

Fig. 5.11: The figure shows the dataset and the thick skyline results. The greyed region around the skyline points (circle and sqaure) indicate the ε neighborhood of those points. Points within those regions are included in the result set for the thick skyline operator.

5.2.3 Constrained Skyline

The constrained skyline operator returns the objects that are part of the skyline when assessed on the subset of the data that satisfy the constraints.

$$ConsSkyline(q, \mathcal{X}) = Skyline(q, \{x | x \in \mathcal{X} \wedge Satisfies(x,C)\}) \qquad (5.22)$$

The construction of the inner dataset for the simple skyline query chooses those objects in $\mathcal{X}$ that satisfy the constraints in the constraint set C. Constraints are often defined as restrictions on values; for example, a constraint may restrict the value of the age attribute on a dataset of people to be between 25 and 35. This operator was proposed in [18]. The constrained skyline result set is different from simply running the basic skyline query and choosing the objects among the skyline set that satisfy the constraint; this is so since the constraints could exclude some objects that had previously dominated objects satisfying the constraints (for simple skyline processing), thus possibly allowing the result set to expand to include the latter. We will see how an object not in the basic skyline could find a place in the constrained skyline, in the example below.

Example: Figure 5.12 shows the result set of the constrained skyline query, where the constraints select only the data points in the greyed region. With the circle and triangle points now falling outside the chosen region, all the dominators for the diamond object are excluded from the search. Thus, this brings the diamond object (which was not a skyline point for the simple skyline query) within the result set for the constrained skyline query.

Motivation: Constraints are useful for scenarios where we would like to restrict the search to only objects that satisfy some property. For example, while matching a

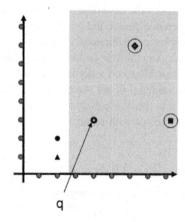

Fig. 5.12: The figure shows the result set of the constrained skyline query where the constraints choose only the dataset falling in the shaded region.

job description against applicant profiles, one may want to select only candidates from a particular nationality, or satisfying a specific age constraint. The constrained skyline query comes in handy for such scenarios.

Group-by Skyline: Group-by skyline [18] refers to executing multiple constrained skyline queries on disjoint subsets of the data, the subsets defined by distinct value combinations of an attribute (or combinations of attributes). For example, while computing skyline of hotels on the attributes of $\{distance, cost\}$, one might be interested in finding the skyline results for hotels of each star rating among the set $\{3*, 4*, 5*\}$. The group-by skyline in this case will produce three different sets of skyline results, each one comprising hotels of a specific star rating among $\{3*, 4*, 5*\}$. This would help in accounting for star ratings of hotels in a fashion which is different from either (a) accounting for it as a skyline attribute, or (b) excluding it from computing the skyline results.

5.2.4 Dynamic Skyline: Skyline in a Transformed Space

The typical transformation applied to convert the $d(.,.)$ vectors to $d_\mathscr{O}(.,.)$ vectors in the basic skyline operation is the identity transformation. Thus, $d_\mathscr{O}(.,.)$ is exactly the same as the corresponding $d(.,.)$ vector. Dynamic skyline [18] uses an alternative transformation that allows the attribute distances to be combined to generate a smaller length $d_\mathscr{O}(.,.)$ vector. The skyline operation then done exactly in the same manner as in the case of the regular skyline operation. Due to the simplicity of this operator, we do not attempt to develop a declarative representation of the results.

Motivating Example: Consider the case of a search for a hotel, where the ideal hotel is one close to the conference center and is very inexpensive. For simplicity, we will assume that the geographic co-ordinates for a hotel is represented by the x and y co-ordinates on a grid rather than latitude and longitude. The third attribute of interest is the cost. While it is possible to do a skyline on the three-attribute space, what is critical to the choice of the hotel is not the combination of x and y co-ordinates of the hotel, but the distance to the conference center that could be approximated by the L_2 norm of the $(\delta x, \delta y)$ vector with respect to the query. Consider the example $d(.,.)$ vector below:

$$d(q,x) = [\delta x = 3.0, \delta y = 4.0, \delta cost = 10.0] \tag{5.23}$$

Under the transformation where the distance would be approximated by the L_2 norm, the transformed vector would be the following:

$$d_\theta(q,x) = [\delta dis = 5.0(= \sqrt{3^2 + 4^2}), \delta cost = 10.0] \tag{5.24}$$

where δdis is the result of the transformation using the δx and δy elements. The skyline results would then be arrived at by processing the transformed vectors instead of the original ones.

5.2.5 Skyband: At most k dominators

The skyband operator relaxes the skyline condition to include all objects that are dominated by at most k objects in the result set.

$$Skyband(q, \mathcal{X}) = \{x | x \in \mathcal{X} \wedge |dominators(x, q, \mathcal{X})| \leq k\} \tag{5.25}$$

where $dominators(x, q, \mathcal{X})$ is the set of objects in $\mathcal{X}$ that dominate x with respect to the query q. Formally:

$$dominators(x, q, \mathcal{X}) = \{y | y \in \mathcal{X} \wedge y \succ_q x\} \tag{5.26}$$

It may be noted that the skyband set reduces to the simple skyline set when $k = 0$. Also, with increasing values for k, the skyband result set monotonically increases as is evident from the construction of the result set.

Example: Figure 5.13 shows the results of the skyband operator for $k = 1$. While the basic skyline objects, the circle and the square remain in the skyband, the triangle also finds place in the skyband set since it is dominated only by one object, the circle. The diamond object, due to having two dominators, would only start being part of the skyband result from $k = 2$ upwards.

Motivation: Skyband is motivated by scenarios similar to that described in the case of the thick skyline operator. In cases where the basic skyline results are unusable due to factors external to the search system (e.g., room unavailability in a hotel in the context of a hotel search system), the skyband provides additional results that

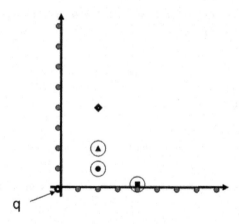

Fig. 5.13: The figure shows the skyband result set for $k = 1$ using a plot of the $d_{\mathcal{O}}(.,.)$ vectors. The triangle object also is included in the skyband since it is dominated only by one object, i.e., the circle. The diamond object, on the other hand, is dominated by both the circle and triangle objects.

the user could fall back on. Thus, thick skyline and skyband operators propose different methods to provide results in addition to those provided by the basic skyline operator.

5.2.6 Reverse Skyband: Indirection on the Skyband

The Reverse k Skyband (*RkSB*) operator [16], as the name implies, refers to the usage of the indirection feature on the skyband operator. Accordingly, the result set is specified as the following:

$$RkSB(q, \mathcal{X}) = \{x | x \in \mathcal{X} \wedge q \in Skyband(x, \mathcal{X} \cup \{q\} - \{x\})\} \qquad (5.27)$$

This operator is a simple extension of the reverse skyline operator where the indirection is performed on the skyband instead of the skyline. Having discussed reverse skyline and other indirected operators earlier, we hope that the semantics of the *RkSB* operator would be quite clear. As in the case of any other scenario where indirection is used, this is helpful in quantifying the influence or importance of an object as well as to discriminate between resources in a resiliency planning scenario.

5.2.7 K-Most Representative Skyline: Maximizing combined dominating Power

The K-Most Representative Skyline operator produces a result set of exactly k objects in such a way that the total number of distinct objects in $\mathscr{X}$ that are dominated by one of the results is maximized:

$$KMRSkyline(q, \mathscr{X}) = \underset{X \subseteq \mathscr{X} \wedge |X| = k}{argmax} \; |\cup_{x \in X} dominated(x, q, \mathscr{X})| \qquad (5.28)$$

where $dominated(x, q, \mathscr{X})$ denotes the set of objects in $\mathscr{X}$ that are dominated by x with respect to the query q:

$$dominated(x, q, \mathscr{X}) = \{y | y \in \mathscr{X} \wedge x \succ_q y\} \qquad (5.29)$$

This is analogous to the definition of $dominators(.,.,.)$ in the earlier section. The skyline points together dominate each of the remaining objects in the dataset; if this were not true, those non-dominated points would have to be be part of the skyline, invalidating the assertion that we started with the set of skyline points. In this context, the *KMRSkyline* is useful in cases when the Skyline result set is large since it enables one to choose a k sized subset of the skyline that have a large dominating power. On the other hand, if the Skyline result set already has fewer than k elements, any choice of additional non-skyline elements to enlarge the *KMRSkyline* set to k would be random choices since none of the non-skyline points can increase the total number of dominated points. The *KMRSkyline* operator was proposed in [15].

Example: Figure 5.14 illustrates the result set of the *KMRSkyline* operator using a plot of $d_{\mathscr{O}}(.,.)$ vectors. Since only one object can be chosen in the result set, the circle object is the obvious choice since it dominates two other objects whereas the other skyline object, the square, is seen to not have any objects in its *dominated* set.

Motivation: Since the result set of the skyline operator is not bounded in cardinality, it is likely that the skyline operator may return plenty of objects as results in certain scenarios. However, due to reasons such as limited screen size in the device on which the result set needs to be displayed, or to enable the user to sift through the result set quickly, we may need to restrict the result set to a manageable size. The simplest way to do that is to pre-specify the number of results required, as in the case of the top-k operator. The *KMRSkyline* operator does precisely that; since the skyline result set is determined by the domination relationship, the *KMRSkyline* operator adapts the notion of domination to the constraint of choosing k results, by choosing the k objects that together dominate the largest number of objects in the dataset.

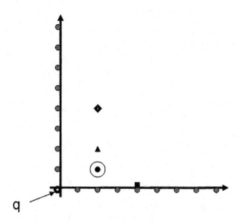

Fig. 5.14: The figure shows the KMRSkyline result set for $k = 1$ using a plot of the $d_{\mathcal{O}}(.,.)$ vectors. Among the skyline objects (i.e., circle and square), the square does not dominate any other non-skyline object whereas the circle dominates both the triangle and the diamond.

5.2.8 Top-k Frequent Skylines: Counting Subspaces

The top-k frequent skylines operator [1] provides another way of choosing a bounded result set while building upon the skyline operator. This operator chooses the k objects in the dataset that have the lowest values for *dominated frequency*[1]:

$$TKFSkyline(q, \mathcal{X}) = \underset{X \subseteq \mathcal{X} \wedge |X| = k}{argmin} \sum_{x \in X} dominated\_f(x, q) \qquad (5.30)$$

where the *dominated frequency* is defined as follows:

$$dominated\_f(x, q) = |\{A | A \in (2^{\mathcal{A}} - \{\phi\}) \wedge x \notin Skyline(q, \Pi_A(\mathcal{X}))\}| \qquad (5.31)$$

Informally, we consider every non-empty subset of attributes from $\mathcal{A}$ where $2^{\mathcal{A}}$ denotes the power set of $\mathcal{A}$. For each such subset A, the skyline set is computed with respect to q over $\mathcal{X}$ projected only on those attributes; this is exactly the projection operation that we saw while discussing the subspaces feature. Finally, $dominated\_f(x, q)$ is the number of attribute-subsets where x does not appear among the skyline result sets. As an example, an object that is very close to the query on each attribute in $\mathcal{A}$ is likely to be in the skyline set for most attribute subsets, and thus would end up with a small value for $dominated\_f(.,.)$.

---

[1] We use *dominated* frequency instead of *dominating* frequency as is used in the original paper [1] since we feel that is more consistent with the terminology already introduced in previous sections.

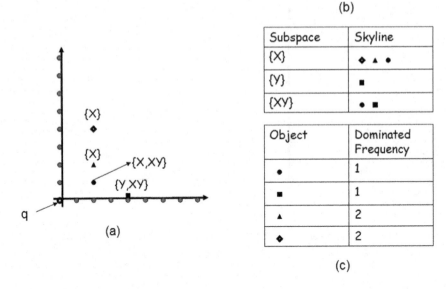

(b)

Subspace	Skyline
{X}	◆ ▲ ●
{Y}	■
{XY}	● ■

Object	Dominated Frequency
●	1
■	1
▲	2
◆	2

(c)

Fig. 5.15: (a) shows the plot of $d_\mathcal{O}(.,.)$ vectors with each data object labeled with the subspaces on which it is in the skyline set. (b) lists the data objects that are part of the skyline set for each subspace. (c) enumerates the dominated frequencies for each data object.

Example: In our running example with two dimensions, there are three non-null attribute subsets, $\{X\}$, $\{Y\}$ and $\{X,Y\}$; for simplicity, we will denote these as X, Y and XY respectively. The plot of $d_\mathcal{O}(.,.)$ of our running example, in Figure 5.15 has the data objects annotated with the subsets of attributes on which they are part of the skyline. When it comes to subsets of just one dimension, all the objects at the least distance from the query on that attribute would be part of the skyline. On X, consequently, all three objects, the circle, the triangle and the diamond, become part of the skyline. Figure 5.15(b) shows the skyline object for each subspace. In Figure 5.15(c), we have each data object along with the count of the number of subspaces on which it is not part of the skyline, i.e., the *dominated frequency*. The *TKFSkyline* operator would choose the k data objects having the lowest values in the table, as the result set.

Motivation: The motivation for choosing k results from the skyline is very compelling in cases where the skyline output is too large to fit on a device screen, or to peruse in a reasonable amount of time. The *TKFSkyline* provides a method, different from that of *KMRSkyline*, to restrict the result set to a pre-specified number of objects, i.e., k. [1] shows an example, where performance of basketball players per season chosen according to the scoring on dominated frequency, is intuitively a good choice of all-time greats.

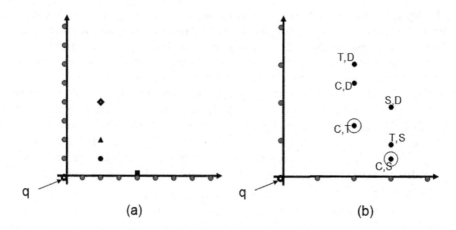

Fig. 5.16: (a) shows the plot of the $d_\Theta(q,x)$ vectors as has been seen earlier. (b) plots the $d_\Theta(q,g)$ vectors where g denotes any 2-element subset of $\mathscr{X}$ using the mean as the aggregation function. Each element is annotated with the components of the group, where the component data objects are denoted by the initials, i.e., C=circle, D=diamond, T=triangle, S=square. The results of the Skyline k-Group operator are circled as usual.

5.2.9 Skyline k-Groups: Group-level Skyline Evaluation

The Skyline k-Group operator [13] searches for skyline results in the space of groups of k objects from $\mathscr{X}$. We will outline the semantics of the operator in a constructive fashion; the first operation is to pre-process the dataset to compute $\mathscr{G}$, the set of distinct k-sized subsets of $\mathscr{X}$.

$$\mathscr{G} = \{g|g \subseteq \mathscr{X} \wedge |g| = k\} \tag{5.32}$$

The skyline k-group operator on $\mathscr{X}$ is simply the skyline operator on $\mathscr{G}$. However, the construction of the $d_\Theta(.,.)$ vectors need to be outlined to complete the semantics of the Skyline operator. This is as follows:

$$d_\Theta(q,g)[j] = agg(\{d_\Theta(q,x)[j]|x \in g\}) \tag{5.33}$$

Thus, each element of the $d_\Theta(q,g)$ for every $g \in \mathscr{G}$ is formed by an aggregation of the corresponding elements of the $d_\Theta(q,x)$ vectors across all elements $x \in g$. The aggregation functions explored in [13] are *average*, *max* and *min*. The final result set of the operator is then specified as follows:

$$SkG(q, \mathscr{X}) = \{g|g \in Skyline(q,\mathscr{G})\} \tag{5.34}$$

It may be noted that the output is a set of elements from $\mathscr{G}$; since each element in $\mathscr{G}$ is a subset of k elements from $\mathscr{X}$, the output of the Skyline k-Group operator is a set of sets. This makes it markedly different from all the operators seen so far.

Example: Figures 5.16(a) and (b) show the $d_\mathcal{O}(q,x)$ and $d_\mathcal{O}(q,g)$ vectors respectively where the latter are computed with $k = 2$. The components of each vector are shown using initials of the component objects, in the latter figure; thus, $[C,T]$ refers to the subset comprising the circle and the triangle data objects. The result set for the Skyline k-Group query in this case happens to be the set $\{\{C,T\},\{C,S\}\}$.

Motivation: In scenarios such as choosing a set of players to function as a team in a sports event, it is not only necessary that the individuals in the team are good; the combined statistics of the team also need to be good enough, to ensure good performance. Skyline k-Group operator characterizes the combined statistics of groups of objects by using one of *average*, *max* or *min* aggregation functions. Thus, the result provides multiple options of k-member teams that could be selected by virtue of their good aggregate profile.

5.2.10 Summary

The skyline operator, as we have seen, has been leveraged in many ways to address scenarios ranging from needs to expand the skyline set (e.g., skyband), restrict the skyline set (e.g., *TKFSkyline*) as well as to channel the search to transformed (e.g., Dynamic Skyline) and potentially expanded (e.g., Skyline k-Groups) spaces. In addition to this, the plain skyline operator has been shown to be useful in emerging scenarios such as location based social networks. Geo-social Skylines [5] are applications of skylines in the space of two attributes, *social distance* and *geographic distance*, to help travelers find more familiar hosts in scenarios like couchsurfing. With the advent of newer forms of data and needs for newer kinds of search mechanisms, it is likely that skyline would continue to be used in unforseen ways in the near future.

5.3 Other Operators

We will now discuss operators that do not use either the skyline or weighted sum framework, or address scenarios that do not fall under either of them. We will also discuss a recent work that outlines novel methods to estimate pair-wise object similarities.

5.3.1 Reverse K-Ranks and Reverse K-Scores

The Reverse K-Ranks query picks the k data objects with respect to whom the query is ranked highest. Formally,

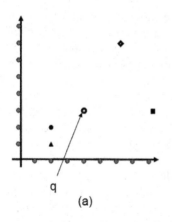

Object	Better(.,.,.)	$d_{WS}(.,.)$
◆	0	6.0
■	0	4.0
●	1	3.0
▲	1	4.0

(a)   (b)

Fig. 5.17: (a) plots a dataset with four data objects along with a query. (b) lists the scores used for data objects in the *Rk-Ranks* and *Rk-Scores* respectively; the distances are computed using a weight vector of all elements set to 1.0.

$$Rk\text{-}Ranks(q, \mathcal{X}) = \underset{X \subseteq \mathcal{X} \wedge |X|=k}{argmin} \sum_{x \in X} better(x, q, \mathcal{X}) \qquad (5.35)$$

where $better(x, q, \mathcal{X})$ counts the number of objects in $\mathcal{X}$ that are more similar to x than the query q is to x:

$$better(x, q, \mathcal{X}) = |\{y | y \in \mathcal{X} \wedge s_{\mathcal{O}=WS}(x,y) > s_{\mathcal{O}=WS}(x,q)\}| \qquad (5.36)$$

where $s_{\mathcal{O}=WS}(.,.)$ is the weighted sum similarity score. $better(x, q, \mathcal{X})$ counts the number of objects in $\mathcal{X}$ that are more similar to x than the query is, to x, and is thus equivalent to the rank of q for a weighted sum query issued on x. The Reverse k-Scores query is similar, but simpler in specification:

$$Rk\text{-}Scores(q, \mathcal{X}) = \underset{X \subseteq \mathcal{X} \wedge |X|=k}{argmax} \sum_{x \in X} s_{\mathcal{O}=WS}(x,q) \qquad (5.37)$$

Thus, the *Rk-Scores* result set contains the set of k objects such that the query object is judged highly similar with respect to them. It may be noted that if the distance measure is symmetric, i.e., the $s_{\mathcal{O}=WS}(x,q) = s_{\mathcal{O}=WS}(q,x)$, this reduces to simply the weighted sum top-k query. However, in certain scenarios, the weight vector used to compute the $s_{\mathcal{O}=WS}(.,.)$ is dependent on the query; in such cases, the *Rk-Scores* operator has a different semantics than simple top-k. Both the *Rk-Ranks* and *Rk-Scores* operators were proposed in [27].

Example: Figure 5.17(a) simply plots a dataset of four objects along with a query. Figure 5.17(b) lists measures of interest, on a per-object basis, for the *Rk-Ranks* and *Rk-Scores* queries respectively. It may be noted that the last column shows the L_1 distance of each object to the query. The diamond and square objects get a $better(.,.,.)$ score of 0 since all other data objects are farther from them than the query. Under these scores, for $k = 2$, they would be chosen in the *Rk-Ranks* result set. On the other hand, for the *Rk-Scores* query, the top-2 objects would be the circle and triangle owing to the lower distance (i.e., high similarity) to the query. This illustrates the difference in semantics between the two operators that are fairly similar in their result set specifications.

Motivation: Consider the bichromatic *RkNN* query that seeks to find prospective customers for a product, for a scenario such as selecting customers to send targeted offers to. Due to the popularity of common products that are bought frequently by many customers, it is likely to be able to find many prospective customers for popular products. On the other hand, for niche products, the *RkNN* set is likely to be a null or very sparse set even for fairly high values of k. For a store that wants to test the market for such a product, it is useful to find the set of customers who are most likely to buy the niche product, even if they are likely to prefer many popular products ahead of the niche one. *Rk-Ranks* and *Rk-Scores* address ways to find results for such cases where the *RkNN* result set is very sparse even for high values of k.

5.3.2 Spatial Skyline

The spatial skyline query [20] explores an interesting combination of skyline and weighted sum operators under a multi-query scenario. Consider a set of query points $Q = \{q_1, q_2, \ldots\}$; now, we will define the operator-specific distance vectors using the distances of each object to each of the query points.

$$d_\mathcal{O}(x, Q) = [d_{\mathcal{O}=WS}(x, q_1), d_{\mathcal{O}=WS}(x, q_2), \ldots] \qquad (5.38)$$

Thus, the $d_\mathcal{O}(x, Q)$ vector has as many entries as there are queries in Q, with the ith entry being the distance to the ith query object. These vectors may be thought of as a transformed representation of the dataset with respect to the query object; we will denote the set of these vectors as a dataset $\mathcal{D}_{Q\mathcal{X}}$:

$$\mathcal{D}_{Q\mathcal{X}} = \{d_\mathcal{O}(x, Q) | x \in \mathcal{X}\} \qquad (5.39)$$

Now, the spatial skyline is simply the skyline of these vectors with respect to a query point having all zeros:

$$SpatialSkyline(Q, \mathcal{X}) = Skyline([0.0, 0.0, \ldots], \mathcal{D}_{Q\mathcal{X}}) \qquad (5.40)$$

where the distance vectors for the inner skyline queries are formed by calculating the numerical difference to the all-zero vector which is equivalent to the component object vectors in $\mathcal{D}_{Q\mathcal{X}}$ themselves. Thus, the results of the spatial skyline query are formed by the data objects that are in the skyline when the skyline is computed using the distance vector to the query objects. Since the component operators, skyline and weighted sum, as well as the multi-query feature have been discussed in detail, we hope that the semantics of the spatial skyline operator is also clear. The metric-space skyline [7] query extends the spatial skyline query to metric spaces.

Motivating Example: Consider the case of a multi-member task force who work from separate offices; they wish to shortlist a list of restaurants for periodic lunch meetings. These locations must be interesting in terms of traveling distances for *all* the team members. In particular, given two restaurants where one is at most as far as the other for each member in the team, the team would prefer the former restaurant, since it is at least as good as the latter for *all* team members and better for *some* team members. This is exactly the skyline result on the distance vectors, i.e., the semantics of the spatial skyline query. Thus, the set of desired restaurants may be shortlisted by running a spatial skyline query on a restaurant database using the locations of each team member as a query.

5.3.3 Hypermatching

We now examine a recent work on similarity matching [25] that considers several deviations from the weighted sum similarity estimation model. We will look at the many deviations presented in the work, one at a time.

Object-specific Weight Vectors: Firstly, the work considers similarity estimation when weight vectors for attribute weighting are provided on a per-object basis. Under the single weight vector scheme that we have considered till now, the similarity between a pair of objects (q, x) would be estimated as:

$$\sum_{1 \leq i \leq m} w_i \times s(q, x)[i] \tag{5.41}$$

However, when each of q and x have separate weight vectors w^q and w^x respectively, [25] proposes that a single weight vector w' be constructed using the following formula:

$$w'_i = \frac{max\{w_i^q, w_i^x\}}{\sum_{1 \leq j \leq m} max\{w_i^q, w_i^x\}} \tag{5.42}$$

The w' vector is constructed by taking the largest value among the two vectors (i.e, w^q and w^x) in each of the positions, followed by normalizing so that the elements sum up to 1.0. The similarity between the objects may be then computed by replacing the weight vector w by this derived weight vector w'.

Different Attribute Sets for Objects: The second deviation considered is about handling cases where the two objects need not necessarily have the same set of attributes. This is a significant departure from the schema-based object representation, but, has parallels in other areas such as feature-based object representation (e.g., Tversky Index[2]). Let us consider $\mathscr{A}$ as a shared attribute space; q and x take values for different subsets of attributes from $\mathscr{A}$. Let w^q and w^x be the initial weight

---

[2] http://en.wikipedia.org/wiki/Tversky_index - Accessed March 27th, 2015

vectors that have weights for all elements in $\mathscr{A}$. We now define a new weight vector $\hat{w}^q$:

$$\hat{w}_i^q = \begin{cases} w_i^q & \text{if } a_i \in q \\ 0.0 & \text{otherwise} \end{cases} \tag{5.43}$$

A similar operation yields $\hat{w}^x$:

$$\hat{w}_i^x = \begin{cases} w_i^x & \text{if } a_i \in x \\ 0.0 & \text{otherwise} \end{cases} \tag{5.44}$$

In each of these cases, the original weight is inherited if the corresponding feature is present; for features that are absent, the weight is set to 0.0. The element-wise max operation is applied as earlier on these two vectors to get a new weight vector $\hat{v}$:

$$\forall i, \hat{v}_i = max\{\hat{w}_i^q, \hat{w}_i^x\} \tag{5.45}$$

This is followed by the computation of the total *potency* of the shared attributes:

$$t(q,x) = sum\{\hat{v}_i | a_i \in q \wedge a_i \in x\} \tag{5.46}$$

We are now ready to formulate the similarity between q and x:

$$s_\mathscr{O}(q,x) = \sum_{i,a_i \in q \wedge a_i \in x} \frac{\hat{v}_i}{\sum_i \hat{v}_i} \times s(q,x)[i] \tag{5.47}$$

Informally, the similarities on only the shared attributes are considered for computation of the similarities. However, since the $\hat{v}$ vector is normalized over all attributes (within the product term), the non-shared attributes have their influence; for example, if the non-shared attributes have a total mass of 0.3 in the $\hat{v}$ vector, the total similarity between q and x can only be at most 0.7 (assuming the $s(q,x)[i]$ is in $[0,1]$ for all i).

Amplifying Extreme Values: The third and final notable deviation considered is that of amplifying the importance of extreme values in objects. Intuitively, an extraordinarily obese man is characterized more by his obesity than other features. Thus, if an object has extreme (or unusual) values for certain attributes, it is usually useful to amplify the importance of that attribute in the weight vector for that object. Let us consider the weight vector for an object x as w^x. This weight vector is modified to account for extreme values by the following transformation:

$$w_i^x = \frac{w_i^x \times Ampl(i,x)}{\sum_j w_j^x \times Ampl(j,x)} \tag{5.48}$$

where $Ampl(j,x)$ is directly related to the level of deviation expressed for the attribute a_j in object x. In particular, $Ampl(j,x)$ would be in the range $[1,\infty)$, with 1 being the value assigned for the case where the value of a_j in x is the most typical value, and very high values reserved for extremely unusual values. Such modified weight vectors may then be used in similarity assesment as in the first and second cases.

Remarks: The work under discussion outlines multiple novel ways of computing object pair-wise similarities. Such estimates may be used as the s_θ values in operators such as top-k, range, *RkNN* etc. Thus, hypermatching, unlike most previously discussed work, is not an operator, but, lays down a method for estimating similarities which could be used to create variants of multiple operators. It could be thought of as another aggregation function and all the features applicable to weighted sum could also be applied to it to generate operators based on hypermatching.

5.3.4 Summary

In this section, we discussed a new way, called *Hypermatching*, of estimating pair-wise object similarity that do not fall under the weighted sum model methods. We also saw a novel operator that enables address sparsity in *RkNN* results, as well as the spatial skyline operator. There have been several other variants of similarity search operators, and similarity estimation, that have been explored in literature, though not extensively. A notable example is the generalization of queries to those having a spatial extent [19]; this helps perform similarity search under queries where a range of values is specified on certain attributes in the query. In particular, the Range Reverse Nearest Neighbor query finds the set of objects that have any point in the query range as their nearest neighbor.

Operator	Section	Operator	Section
Weighted Sum Top-k	5.1	Weighted Sum Range	5.1
RkNN	5.1.1	BRkNN	5.1.1
RkFN	5.1.3	Constrained NN	5.1.4
Constrained kNN	5.1.4	Visible kNN	5.1.5
Obstacle NN	5.1.4	Subspace Top-k	5.1.6
Subspace Range	5.1.6	KNDN	5.1.7
Multi-query Top-k	5.1.8	Skyline	5.2
Reverse Skyline	5.2.1	Thick Skyline	5.2.2
Constrained Skyline	5.2.3	Group-by Skyline	5.2.3
Dynamic Skyline	5.2.4	Skyband	5.2.5
Reverse Skyband	5.2.6	K-Most Representative Skyline	5.2.7
Top-k Frequent Skylines	5.2.8	Skyline k-Groups	5.2.9
Geo-social Skyline	5.2.10	Reverse k-Ranks	5.3.1
Reverse k-Scores	5.3.1	Spatial Skyline	5.3.2
Hypermatching Formulations	5.3.3	Range Reverse Nearest Neighbor	5.3.4

Table 5.12: Section-wise Summary of Similarity Operators

5.4 Summary

In this chapter, we considered various advanced operators for similarity search under three heads, (a) those that build upon the weighted sum operation (or any scalar aggregation function), (b) operators that enhance the basic skyline operator and (c) other modes of similarity search not already covered under the first two heads. Table 5.12 lists the set of operators that have been discussed in this section along with the section names, as a ready reference. In the context of newer kinds of data from emerging sources like social networks, GPS-traces, user logs and so on, it is reasonable to expect that research on devising new operators would continue at an even faster pace in the years to come.

References

1. C.-Y. Chan, H. Jagadish, K.-L. Tan, A. K. Tung, and Z. Zhang. On high dimensional skylines. In *Advances in Database Technology-EDBT 2006*, pages 478–495. Springer, 2006.
2. C. L. Clarke, M. Kolla, G. V. Cormack, O. Vechtomova, A. Ashkan, S. Büttcher, and I. MacKinnon. Novelty and diversity in information retrieval evaluation. In *Proceedings of the 31st annual international ACM SIGIR conference on Research and development in information retrieval*, pages 659–666. ACM, 2008.
3. E. Dellis and B. Seeger. Efficient computation of reverse skyline queries. In *Proceedings of the 33rd international conference on Very large data bases*, pages 291–302. VLDB Endowment, 2007.
4. P. M. Deshpande and D. Padmanabhan. Efficient reverse skyline retrieval with arbitrary non-metric similarity measures. In *EDBT 2011, 14th International Conference on Extending Database Technology, Uppsala, Sweden, March 21-24, 2011, Proceedings*, pages 319–330, 2011.
5. T. Emrich, M. Franzke, N. Mamoulis, M. Renz, and A. Züfle. Geo-social skyline queries. In *Database Systems for Advanced Applications*, pages 77–91. Springer, 2014.
6. H. Ferhatosmanoglu, I. Stanoi, D. Agrawal, and A. El Abbadi. Constrained nearest neighbor queries. In *Advances in Spatial and Temporal Databases*, pages 257–276. Springer, 2001.
7. D. Fuhry, R. Jin, and D. Zhang. Efficient skyline computation in metric space. In *Proceedings of the 12th International Conference on Extending Database Technology: Advances in Database Technology*, pages 1042–1051. ACM, 2009.
8. Y. Gao, B. Zheng, G. Chen, W.-C. Lee, K. C. Lee, and Q. Li. Visible reverse k-nearest neighbor queries. In *Data Engineering, 2009. ICDE'09. IEEE 25th International Conference on*, pages 1203–1206. IEEE, 2009.
9. A. Jain, P. Sarda, and J. R. Haritsa. Providing diversity in k-nearest neighbor query results. In *Advances in Knowledge Discovery and Data Mining*, pages 404–413. Springer, 2004.
10. W. Jin, J. Han, and M. Ester. Mining thick skylines over large databases. In *Knowledge Discovery in Databases: PKDD 2004*, pages 255–266. Springer, 2004.
11. F. Korn and S. Muthukrishnan. Influence sets based on reverse nearest neighbor queries. In *ACM SIGMOD Record*, volume 29, pages 201–212. ACM, 2000.
12. Y. Kumar, R. Janardan, and P. Gupta. Efficient algorithms for reverse proximity query problems. In *Proceedings of the 16th ACM SIGSPATIAL international conference on Advances in geographic information systems*, page 39. ACM, 2008.
13. C. Li, N. Zhang, N. Hassan, S. Rajasekaran, and G. Das. On skyline groups. In *Proceedings of the 21st ACM international conference on Information and knowledge management*, pages 2119–2123. ACM, 2012.

14. X. Lian and L. Chen. Similarity search in arbitrary subspaces under l p-norm. In *Data Engineering, 2008. ICDE 2008. IEEE 24th International Conference on*, pages 317–326. IEEE, 2008.

15. X. Lin, Y. Yuan, Q. Zhang, and Y. Zhang. Selecting stars: The k most representative skyline operator. In *Data Engineering, 2007. ICDE 2007. IEEE 23rd International Conference on*, pages 86–95. IEEE, 2007.

16. Q. Liu, Y. Gao, G. Chen, Q. Li, and T. Jiang. On efficient reverse k-skyband query processing. In *Database Systems for Advanced Applications*, pages 544–559. Springer, 2012.

17. S. Nutanong, E. Tanin, and R. Zhang. Visible nearest neighbor queries. In *Advances in Databases: Concepts, Systems and Applications*, pages 876–883. Springer, 2007.

18. D. Papadias, Y. Tao, G. Fu, and B. Seeger. Progressive skyline computation in database systems. *ACM Transactions on Database Systems (TODS)*, 30(1):41–82, 2005.

19. R. Pereira, A. Agshikar, G. Agarwal, and P. Keni. Range reverse nearest neighbor queries. In *KICSS*, 2013.

20. M. Sharifzadeh and C. Shahabi. The spatial skyline queries. In *Proceedings of the 32nd international conference on Very large data bases*, pages 751–762. VLDB Endowment, 2006.

21. Y. Shi and B. Graham. A similarity search approach to solving the multi-query problems. In *Computer and Information Science (ICIS), 2012 IEEE/ACIS 11th International Conference on*, pages 237–242. IEEE, 2012.

22. Y. Tao, D. Papadias, and X. Lian. Reverse knn search in arbitrary dimensionality. In *Proceedings of the Thirtieth international conference on Very large data bases-Volume 30*, pages 744–755. VLDB Endowment, 2004.

23. Y. Tao, X. Xiao, and J. Pei. Efficient skyline and top-k retrieval in subspaces. *Knowledge and Data Engineering, IEEE Transactions on*, 19(8):1072–1088, 2007.

24. A. K. Tung, R. Zhang, N. Koudas, and B. C. Ooi. Similarity search: a matching based approach. In *Proceedings of the 32nd international conference on Very large data bases*, pages 631–642. VLDB Endowment, 2006.

25. R. Yager and F. Petry. Hypermatching: Similarity matching with extreme values. *Fuzzy Systems, IEEE Transactions on*, 22(4):949–957, Aug 2014.

26. J. Zhang, D. Papadias, K. Mouratidis, and M. Zhu. Spatial queries in the presence of obstacles. In *Advances in Database Technology-EDBT 2004*, pages 366–384. Springer, 2004.

27. Z. Zhang, C. Jin, and Q. Kang. Reverse k-ranks query. *Proceedings of the VLDB Endowment*, 7(10), 2014.

Chapter 6
Indexing for Similarity Search Operators

In the previous chapters, we have seen that there is a plethora of similarity search operators that have been proposed till date. The natural question that needs to be addressed is how to run the various similarity search operators efficiently. Similarity search systems operate on large datasets that cannot be held in memory or traversed in entirety for each query. Thus, indexes and efficient algorithms to process such operators are a prerequisite to making operators usable in practical scenarios. Indexing methods have been well-studied with many focused surveys and books [30, 23, 3]. Our goal is not to give a comprehensive survey of the indexing techniques used for similarity search. Rather, we intend to give only a flavor of the indexing techniques and algorithms used for similarity search operators. The indexing techniques depend heavily on the properties of the distance measures and the similarity operators that need to be supported. We take top-k as the representative similarity query and outline one or two such indexes for each scenario — distance measures in an Euclidean space, metric distances and non-metric distances. In each scenario, we also briefly outline the algorithm for processing the top-k query using such indexes.

6.1 Euclidean Space Indexes

By distances being in Euclidean space, we mean that the data objects can be mapped to points in a n-dimensional Euclidean space, such that the distance between the objects on each attribute corresponds to the distance between the corresponding points on the corresponding dimension. There have been many indexes proposed in literature for querying of spatial data in a Euclidean space, such as R-Tree [15], k-d Tree [2], Quadtree [12], Grid index [20], etc. These indexes can be used for many types of similarity queries as well. Since these indexes are well known, we will not describe them in detail. To give an idea of how they can be used for similarity search, we will cover one example below.

© The Author(s) 2015
D.P and P.M. Deshpande, *Operators for Similarity Search,*
SpringerBriefs in Computer Science, DOI 10.1007/978-3-319-21257-9_6

6.1.1 Top-k using k-d Trees

The k-d Tree is a space partitioning index that can be considered as a multi-dimensional binary tree. Each node in the tree splits the space into two based on the values along one of the dimensions. All nodes at the same level split the space on the same dimension. The dimension chosen for splitting keeps cycling between the dimensions as we move down the levels of the tree. Ideally each split partitions the sub-space into two parts such that there are equal number of points in each part. This is achieved by ordering the points along the splitting dimension and choosing the dimension value of the median point as the splitting value. For example, consider the data points shown in Figure 6.1(a). We start with A_1 as the splitting dimension at the root. Ordering the points on the A_1 attribute, we can see that $X_7 : (3,2)$ is in the middle with 4 points to the left with $A_1 < 3$ and 3 points to the right with $A_1 > 3$. Thus, X_7 is the splitting point and A_1 is the dimension to split on. Next, we choose A_2 as the splitting dimension to further split each half of the space to the left and right of the root. In the left sub-space, $X_3 : (1,4)$ becomes the split point with A_2 being the dimension to split on. Similarly, in the right sub-space, $X_5 : (6,3)$ becomes the point to split on along the A_2 dimension. Completing this process leads to the space partitioning shown in Figure 6.1(a) and the corresponding k-d tree is shown in Figure 6.1(b). Searching, inserts and updates follow a path from the root to the leaf by comparing the corresponding dimension value at each level with the split point and taking the left or right path based on the result. For example, consider the search for $(5,1)$. At the root node, we will take the path to the right child, since the value on A_1 attribute of 5 is greater than the value 3 for the root node. Then, at the node X_5, we take the path to the left child since the value on A_2 of 1 is less than the value 3 of X_5 on A_2. This leads us to the leaf node corresponding to X_8 which has the attribute values $(5,1)$. Inserts and deletes could lead to an unbalanced tree. There is no easy way to re-balance the tree after each operation. The entire tree needs to be re-created periodically when it gets very skewed.

Let us now look at how to perform top-k search using the k-d Tree. For simplicity we will consider the top-1 (nearest neighbor) search. The algorithm has the following steps:

1. Traverse the tree to find the leaf(l) corresponding to the query point q. Initialize *current_best* to l and set d_{best} to be $d_{\mathcal{O}}(q,l)$.
2. Traverse upwards towards the root. For each node n on the path

   - If $(d_{\mathcal{O}}(q,n) < d_{best})$ update *current_best* to n and set d_{best} to be $d_{\mathcal{O}}(q,n)$.
   - Let i be the dimension on which the node n splits the space
   - If $(d(q,n)[i] <= d_{best})$ explore the other sub-tree of n using the same recursive procedure updating *current_best* and d_{best} in the process

At each node, the algorithm checks if the other sub-tree needs to explored by checking if there can be any points on the other side of the dividing plane that are closer than the current best point. This is done by just checking the closest distance on the splitting dimension. If the closest distance on the splitting dimension $d(q,n)[i]$

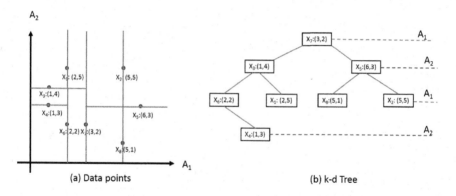

Fig. 6.1: k-d Tree Construction

is less than d_{best} then there is a possibility of finding a nearest point on the other side and so the other sub-tree must be explored. It can be seen that this condition works for monotonic distance functions $d_\mathcal{O}$. Consider the same example as before with $q : (5, 2.5)$ as the query point as shown in Figure 6.2. The sequence of nodes visited is shown in Figure 6.2(b). The table at the top shows the current best node and its distance at various points in the sequence. The initial search with q leads to the leaf X_8 (steps 1 and 2). This sets current best as X_8 and the distance as 1.5 which is the distance between X_8 and q. Then the algorithm starts traversing upwards to the root. At step 3, X_5 is closer than the current best. So, current best is set to X_5 and the distance is set to 1.11, which is the euclidean distance between X_5 and q. Further, A_2 is the splitting dimension at X_5. The distance between q and X_5 on A_2 is 0.5 which is less than 1.11, so there is a potential nearer point on the other side (see circle in Figure 6.2(a)). So the right subtree is explored in steps 4 and 5, but the current best does not get updated. Finally, we reach the root X_7 in step 6. At the root, A_1 is the splitting dimension. The distance between X_7 and q on A_1 is 2 which is larger than 1.11. So the entire subtree on the left of the root need not be explored at all. The procedure for top-k is similar, the only difference being that instead of maintaining just one best point, we maintain the current top-k set and the farthest distance to any point in that set.

6.2 Metric Space Indexes

In cases where the data cannot be mapped to an Euclidean space, the distance function d could still be a metric. Metric spaces are those where the distances satisfy the metric properties, which can be listed as follows:

1. $d(x_i, x_j) \geq 0$, $d(x_i, x_i) = 0$ (non-negative)
2. $d(x_i, x_j) = d(x_j, x_i)$ (symmetric)

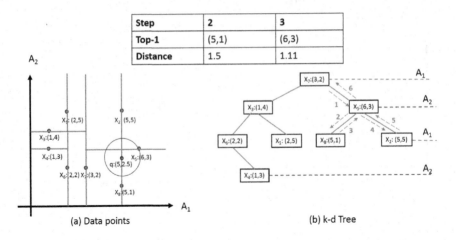

Step	2	3
Top-1	(5,1)	(6,3)
Distance	1.5	1.11

(a) Data points   (b) k-d Tree

Fig. 6.2: Top-k using k-d Tree

3. $d(x_i,x_k) \le d(x_i,x_j) + d(x_j,x_k)$ (triangle inequality)

The most useful property that can be used in similarity search is the triangle in-equality. Metric space indexes aim to exploit this property to prune parts of the space during the search process. Several indexes have been proposed over time in-cluding the Vantage point tree or metric tree [26, 29], D-index [8], M-tree [5] and others [27]. Note that Euclidean spaces are special cases of metric spaces, since the Euclidean distance satisfies the metric properties. Several books [30, 23] provide a comprehensive surveys of metric space indexing. Here, we will describe the Vantage point or VP-Tree as an example of metric space indexing.

For multi-attribute objects, either the distances on each attribute or the aggregate distance $d_\mathcal{O}$ or both could be metric. The common approach is to consider the ag-gregate distance $d_\mathcal{O}$ to be a metric, so we will use the same assumption in describing the metric based indexing methods. If only some of the attribute distances are met-ric, but not the aggregate distance, these methods could still be used by building the metric indexes on the attributes that satisfy the metric properties and performing the aggregation by combining across multiple such indexes.

6.2.1 Top-k using VP-Trees

The VP tree is a space partitioning structure which is based on spherical balls. Each node in the tree represents a sphere with the center as a data point x (called the van-tage point) and a radius r. The remaining points in the data space are partitioned into two – those that are at a distance less than r from x and those that are at a distance greater than r from x. These two partitions are mapped into the left and right sub-

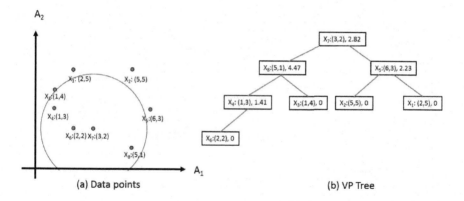

Fig. 6.3: VP Tree Construction

trees of the node respectively. The left and right sets are recursively partitioned in a similar fashion to create the left and right sub-trees. The only information needed to create the VP trees are the distances from the vantage point, which should satisfy metric properties. The individual dimension values are not needed. A simple strategy to construct the tree proceeds as follows. Choose a data point randomly as a vantage point, say x. Compute distances of all the other points from x. Find the median distance and use it as the radius r. Create a node (x, r) and partition the rest of the points into two sets L and R based on whether their distance from x is less than or greater than r. Recursively create the trees for L and R and add them as the left and right sub-trees for the node (x, r). Consider the example shown in Figure 6.3. Initially X_7 is chosen as the vantage point. Computing the distances of the remaining points from X_7 and choosing the median distance, splits the points into two sets: $\{X_3, X_4, X_6, X_8\}$, which are at a distance less than 2.82 from X_7 and $\{X_1, X_2, X_5\}$, which are at a distance greater than 2.82. Note that 2.82 is the median distance (given by point X_3) of the points from X_7. The root node is thus $(X_7, 2.82)$. Each of these sets is similarly recursively divided leading to the tree shown in Figure 6.3(b). Searching in the VP-tree is fairly straightforward. Starting from the root, we compute the distance of the search point from the vantage point. If the distance is less than the threshold distance, we descend down the left sub-tree, otherwise, we descend down the right sub-tree.

Let us now look at Top-k computation using the VP-Tree. The pruning due to the VP-Tree is achieved by using the triangle inequality on the distances. Let (v, r) be a node in the tree, where v is the vantage point and r is the threshold distance. Let q be the query point and x denote any point in the right sub-tree, i.e. point outside the circle around v. We can derive the following inequalities:

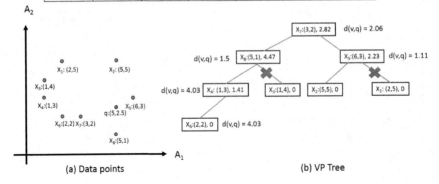

Node	X_7	X_8	X_4	X_6	X_5	X_2
Top-1	(3,2)	(5,1)	(5,1)	(5,1)	(6,3)	(6,3)
Distance	2.06	1.5	1.5	1.5	1.11	1.11

(a) Data points   (b) VP Tree

Fig. 6.4: Top-k using VP Tree

$$d(v,x) \leq d(v,q) + d(q,x), \text{ triangle inequality}$$
$$d(v,x) - d(v,q) \leq d(q,x)$$
$$r - d(v,q) \leq d(q,x), \text{since } r < d(v,x) \text{ for } x \text{ outside the circle}$$

This gives an lower bound on the distance $d(q,x)$ for any x outside the circle. If this is greater than the distance of the current k^{th} best point, the entire right sub-tree can be eliminated since none of them can be in the top k. Consider the example shown in Figure 6.4, which shows the nearest neighbor computation (Top-1) for the query point $q : (5,2.5)$. We do a depth first traversal of the tree, maintaining the current best point and its distance. As we traverse down in the order X_7, X_8, X_4, X_6, the top-1 and the distance get updated to $(5,1)$ and 1.5 as shown in the figure. Now consider the decision of whether to take the right path at X_8. From the above inequality, we have $d(q,x) \geq r - d(v,q) \geq 4.47 - 1.5 \geq 2.97$. Since this is greater than the current best distance of 1.5, there can be no candidate nodes in the right sub-tree we can prune the entire path as shown in the figure. The search then proceeds to X_5, X_2, during which the current best and the distance are updated to $(6,3)$ and 1.11. The right path at X_5 also gets pruned, since $r - d(v,q) = 2.23 - 1.11 = 1.12$, which is greater than the current best of 1.11. Thus the final result is $X_5 : (6,3)$.

6.3 Non-metric Space Indexes

As we saw, metric space indexes exploit the triangle inequality property of the distance function. However, the triangle inequality property is too restrictive to model the (dis)similarities as perceived by humans (for example, [13]). In practice, there are several dissimilarity measures that are non-metric [24]. This is particularly true in the case of similarities over categorical value spaces or complex attribute types. These include the k-median distance that measures the kth most similar portion of the compared objects and partial Hausdorff distance (pHD) for shape based image retrieval. Also, real world algorithms for measuring the distance may be complex based on data analysis, learning distances from the distribution, or domain knowledge [18, 16], which makes them non-metric.

One way to handle non-metric distances is to approximate them with metric distances and use metric indexes. Another common approach in a multi-attribute setting is to exploit properties of the distance aggregation function that transforms $d(.,.)$ to $d_\mathcal{O}(.,.)$. An assumption that is frequently used is that the function is monotonic. Consider two distance vectors d_1 and d_2. The monotonicity property states that if $\forall i, d(q,x_1)[i] \leq d(q,x_2)[i]$, then $d_\mathcal{O}(q,x_1) \leq d_\mathcal{O}(q,x_2)$ where the $d_\mathcal{O}(.,.)$ are the aggregate distances computed from the attribute wise distances. It can be seen that the common aggregation functions such as the weighted sum and the Euclidean distance satisfy the monotonicity property.

We will cover two approaches, one that just exploits the monotonicity property in a middleware setting and another that explicitly builds an index on the data to use in the similarity search.

6.3.1 Middleware Algorithms for Top-k

A popular class of algorithms used in top-k query processing is the family of threshold algorithms (TA) [10]. They maintain precomputed index lists on disk, one for each attribute value, sorted in ascending order of attribute dissimilarity and assume that these lists can be accessed by random accesses or sequential accesses. These were proposed in the context of middleware systems, which perform aggregations of the individual lists maintained on the source systems. Threshold algorithms perform scans over these lists and aggregates dissimilarities across various attributes on the fly, thus maintaining a upper bound for the rank-k result candidate and a lower bound for the scores of the rest of the candidates. They stop processing as and when they have processed enough to reach a condition that the latter is greater than the former, which ensures that the top-k candidates are indeed the top-k closest objects to the query. Different algorithms in the TA family differ in how they schedule the random accesses and sequential accesses over the index lists. The basic TA algorithm can be stated as follows:

1. Read the next item in each list.
2. For each item read, do a random access in the other lists to gather all the attributes of the object corresponding to that item
3. Once all the attributes are known, compute the aggregate distance of the object. If the distance is less than the current k^{th} best object, replace it with this new object in the top-k result set. Update the threshold t to be the distance of kth best object in the result set.

X_1	X_2	X_3	X_4	X_5	X_6	X_7	X_8
(2,5)	(5,5)	(1,4)	(1,3)	(6,3)	(2,2)	(3,2)	(5,1)

Data points

q: (5,2.5)

	1	2	3	4	5	6	7	8	
A_1	X_2:0	X_8:0	X_5:1	X_7:2	X_1:3	X_6:3	X_3:4	X_4:4	$d(x,q)[1]$
A_2	X_4:0.5	X_5:0.5	X_6:0.5	X_7:0.5	X_3:1.5	X_8:1.5	X_1:2.5	X_2:2.5	$d(x,q)[2]$

Sorted Lists

Step	1	2	3	4
Top-1	X_2	X_5	X_5	X_5
Threshold	2.5	1.1	1.1	1.1
d_{min}	0.5	0.5	1.1	2.06

$$d_o(x,q) = \sqrt{(d(x,q)[1])^2 + (d(x,q)[2])^2}$$

Top-k processing

Fig. 6.5: Top-k using Threshold Algorithm

4. Compute the distance of a virtual object created from the last read items, say d_{min}. If this distance is greater than the threshold t, stop processing else loop back to Step 1.

The stopping condition works due to the monotonicity property of the distance function. Any object that is not yet seen will have a distance on each attribute greater than the max distance seen till now (since the lists are sorted), and thus, its aggregate distance will be greater than d_{min}. Once d_{min} exceeds t, we can be sure that no new objects can be added to the result set and the search can stop.

Consider the example shown in Figure 6.5, which uses the same data points as before, but without any metric assumptions. The table in the middle shows the lists for the two attributes, where each list is sorted in the order of increasing distance from the query object $q : (5,2.5)$ on that attribute. For example, X_2 has a distance 0 from the query on A_1; so it appears first in the A_1 list. Similarly, X_4 with a distance 0.5 on A_2 appears first in the second list, since that is the minimum distance. The aggregate distance function is the Euclidean distance, which is monotonic. The bottom table shows the d_{min} and t values at various steps in the process of searching for the Top-1 object. On reading the first element in each list, we get A_1 distance of X_2 an A_2 distance of X_4. For X_2, we do a random access in the 2nd list to get its A_2 distance as 2.5. Similarly a random access is done in the first list for X_4. Since X_2 and X_4 are fully specified now, we can compute their aggregate distances as 2.5 and 4.03 respectively. Since X_2 is nearer, we add it to the Top-1 result set and update threshold t to 2.5. The lower bound on the distances of unseen objects d_{min} is computed by aggregating the distances read in this step 0 and 0.5, which results in d_{min} being set to 0.5. We continue this process by reading the next items in each list and updating the result set, t and d_{min} in each step. At the end of step 4, d_{min} is greater than the threshold t, so the algorithm terminates and outputs the Top-1 set. It can be seen that only half of each list was scanned in this example.

Early members of the TA family made extensive use of random access (RA) to index entries to resolve missing attribute-level dissimilarities of result candidates. But, for very large index lists with millions of entries that span multiple disk tracks, the resulting random access cost is 50 - 50,000 times higher than the cost of a sorted access (SA). To address this problem, it has been proposed not to use random access [11, 14]. This variant of TA is called NRA (No RA). However, this may make NRA to scan longer parts of index lists in order to discover incomplete information. Therefore, Fagin proposed a combined algorithm (CA) [9] that occasionally and carefully performs RAs for promising candidates which can contribute to major pruning of candidates. Scheduling of sequential and random accesses efficiently is critical for TA style algorithms and has been addressed by many papers in different scenarios [4, 19, 1, 25]. Other extensions have also been studied, such as online versions of the TA algorithm [17] and relaxing the assumption that the aggregation functions are monotonic [28].

6.3.2 Top-k using AL Trees

Let us now look at the approach of building an index for arbitrary non-metric distance measures. One of the methods proposed for this is the Attribute Level Tree (AL Tree) [6], which is applicable for multi-dimensional data and monotonic aggregate functions. It is a simple structure in which each level represents an attribute, nodes in a level represent various values taken by the attribute and leaves represent the data items. The number of levels in the tree is equal to the number of attributes. The values on the path from the root to the leaf indicate the attribute values of the object represented by the leaf. Sometimes, there may be a chain of nodes in the path to a leaf, each of which has has a single child. Such chains are compressed into a single node for efficiency reasons. Figure 6.6 shows the AL Tree constructed for the example dataset with the same data points as before. The tree has two levels, one each for A_1 and A_2. The nodes have been labeled from A to L to enable referring to them in the following writeup. The actual data stored in the nodes has been shown in the boxes. Consider the node G. The path to G includes the node with value 1 for A_1 and value 5 for A_2. This indicates that the data point at G, X_4 has the attribute values $(1,5)$. There is only point with A_1 attribute value of 3 $(X_7 : (3,2))$. So the chain of nodes has been compressed into the node D, as shown in the figure.

AL Tree can be used for efficient Top-k computation, as we will see next. Given a query point q, the nodes at each level in the tree need to be re-ordered in the increasing order of distance from the query q on the attribute corresponding to that level. It may be noted that this re-ordering is not done on disk; instead, some additional information in the form of indirection lists are leveraged so that the algorithm can visit the tree in the order it would visit if it were stored in the re-ordered fashion. The idea is to visit the nodes in the increasing order of the distances. An example of top-

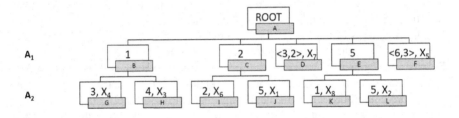

Fig. 6.6: AL Tree

k computation for the query $q : (5, 2.5)$ has been shown in Figure 6.7. At the level A_1, the nodes get re-ordered as E, F, D, C, B. For example, E comes first since its A_1 attribute value of 5 matches the query exactly. Note that the nodes are not actually re-ordered for each query for efficiency reasons. Rather, the same effect is achieved using indirection lists [6]. The search through the AL Tree employs controlled expansion of the tree using the information from the ordering of the siblings in the tree. The progress of the expansion (search) procedure is tracked by maintaining information in the form of 4-tuples, which are referred to hereafter as *candidates*. A candidate points to a node in the AL Tree, and maintains information about which child of the node it would lead the search to, and distance information for the current node. A candidate $C = \, < N, M, d, d_{min} >$ points to a node N, M being the leftmost child of N yet to be explored in the search process. d refers to the distance of the query object to the candidate C based on the attributes seen so far (at each level from the root to N) whereas d_{min} is the lower bound on the distance of any child of C from the query object, based on the AL Tree nodes seen so far. The search procedure works by expanding candidates. The expansion of a candidate leads to visiting a new node, not seen so far, or to the emission of some objects to the result stream. The expansion of C would lead to emission of the objects into the top-k output if N is a leaf node. The expansion of C would yield the following candidates if N is an internal node

1. $< M, P, e, e >$ where $e =$ distance computed using the attribute values seen till now and P is the leftmost child of M AND
2. Either of

    a. $< N, M', d, d' >$ where M' is the right adjacent sibling of M and d' is the distance computed using the attribute value of M OR
    b. *null*, if M is the rightmost child of N

Since the nodes are ordered in the increasing order of distance, any node to the right of M will have a distance d' greater than the distance computed using the attribute value of M. Thus, the d_{min} of any candidate holds the lower bound of distance of any child of the candidate as estimated by the nodes of the tree seen so far. The search procedure maintains a list of candidates, and works by expanding the one in the list with least value for the d_{min}, until at least k nearest neighbors are emitted. The steps are as follows:

1. $S = \{< Root, \text{First child of root}, 0.0, 0.0 >\}$
2. while($S \mathrel{!=} \phi$ AND $|\text{results emitted}| < k$)

    a. Pick P, the candidate in S which has the least value for the d_{min} and expand P
    b. If P is a leaf node, emit the objects that it stands for
    c. Else $S = (S - \{P\}) \cup \{\text{non-null expansions of } P\}$

Consider the example shown in Figure 6.7 again. As before, the distance function is the Euclidean distance, which is monotonic and satisfies the required property. We start by expanding the candidate $< A, E, 0, 0 >$ which indicates the first child of the root node. This leads to $< E, K, 0, 0 >$ which is the first child of E and $< A, F, 0, 0 >$, which indicates the second child of the root. The distances are still 0, since the A_1 attribute value of 5 has matched exactly. In each step, the candidate with the least d_{min} is chosen for expansion. Let us next consider the expansion of $< A, F, 0, 0 >$ in step 3. This leads to two candidates $C_1 =< F, , 1.1, 1.1 >$, which indicates the leaf node F and $C_2 =< A, D, 0, 1 >$, which indicates the next child D of A. The distances in C_1 are 1.1, since that is the distance based on the values seen $(6, 3)$ from the query $(5, 2.5)$ according to the $d_{\mathcal{O}}$ function used. In C_2, d is 0, since no attribute value for D has actually been seen till now. On the other hand, d_{min} is 1 since F has a distance of 1 on the A_1 attribute from the query and any node to the right of F will have at least the same distance due to the order among the nodes. Finally, let us consider the expansion of $< F, , 1.1, 1.1 >$ in step 5. Since this is a leaf node (as indicated by the null child), the objects in it can be directly emitted as a result (in this case, X_5). Since we are interested in only the Top-1, the algorithm need not proceed any further.

As can be seen, this is an online algorithm that emits the results in the increasing order of distance and can stop at any time once the required number of results are found. In general, the AL Tree structure works well for categorical attribute values where there are few distinct values per attribute. It is a versatile structure for non-metric distance measures that can be used for other types similarity queries such as Skyline [22], Reverse Top-k [21] and Reverse Skyline [7] as well.

6.4 Summary

In this chapter, we emphasized the need for indexes and efficient algorithms for evaluating similarity queries. We started with the most favorable case of the points being in a Euclidean space and described a commonly used index the k-d Tree that can be used for efficient search. We then relaxed the constraints and moved to general metric spaces, where the triangle inequality becomes the most important property. We described a commonly used index called the VP Tree for such a case. Finally, we

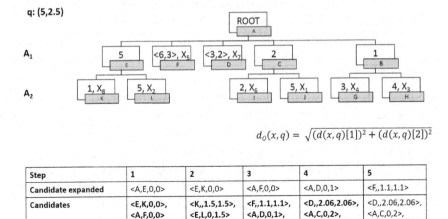

$$d_0(x,q) = \sqrt{(d(x,q)[1])^2 + (d(x,q)[2])^2}$$

Step	1	2	3	4	5
Candidate expanded	<A,E,0,0>	<E,K,0,0>	<A,F,0,0>	<A,D,0,1>	<F,,1.1,1.1>
Candidates	<E,K,0,0>, <A,F,0,0>	<K,,1.5,1.5>, <E,L,0,1.5> <A,F,0,0>	<F,,1.1,1.1>, <A,D,0,1>, <K,,1.5,1.5> <E,L,0,1.5>	<D,,2.06,2.06>, <A,C,0,2>, <F,,1.1,1.1>, <K,,1.5,1.5> <E,L,0,1.5>	<D,,2.06,2.06>, <A,C,0,2>, <K,,1.5,1.5> <E,L,0,1.5>
Output					X_5

Fig. 6.7: Top-k using AL Tree

considered the most general scenario where the distance measure can be arbitrary. The only property assumed is the monotonicity of the distance function. Here we considered two approaches – the family of Threshold algorithms and the AL Tree based indexing method and showed how they can be used for Top-k search in such situations.

References

1. H. Bast, D. Majumdar, R. Schenkel, M. Theobald, and G. Weikum. Io-top-k: Index-access optimized top-k query processing. In *VLDB*, pages 475–486, 2006.
2. J. L. Bentley. Multidimensional binary search trees used for associative searching. *Commun. ACM*, 18(9):509–517, 1975.
3. A. Bhattacharya. *Fundamentals of Database Indexing and Searching*. Taylor & Francis, 2014.
4. K. C.-C. Chang and S. won Hwang. Minimal probing: supporting expensive predicates for top-k queries. In *SIGMOD Conference*, pages 346–357, 2002.
5. P. Ciaccia, M. Patella, and P. Zezula. M-tree: An efficient access method for similarity search in metric spaces. In *VLDB'97, Proceedings of 23rd International Conference on Very Large Data Bases, August 25-29, 1997, Athens, Greece*, pages 426–435, 1997.
6. P. M. Deshpande, P. Deepak, and K. Kummamuru. Efficient online top-k retrieval with arbitrary similarity measures. In *Proceedings of the 11th international conference on Extending database technology: Advances in database technology*, pages 356–367. ACM, 2008.
7. P. M. Deshpande and D. Padmanabhan. Efficient reverse skyline retrieval with arbitrary non-metric similarity measures. In *EDBT 2011, 14th International Conference on Extending*

Database Technology, Uppsala, Sweden, March 21-24, 2011, Proceedings, pages 319–330, 2011.

8. V. Dohnal, C. Gennaro, P. Savino, and P. Zezula. D-index: Distance searching index for metric data sets. *Multimedia Tools Appl.*, 21(1):9–33, 2003.

9. R. Fagin. Combining fuzzy information: an overview. *SIGMOD Record*, 31(2):109–118, 2002.

10. R. Fagin, A. Lotem, and M. Naor. Optimal aggregation algorithms for middleware. *Journal of Computer and System Sciences*, 66(4):614–656, 2003.

11. R. Fagin, A. Lotem, and M. Naor. Optimal aggregation algorithms for middleware. *J. Comput. Syst. Sci.*, 66(4):614–656, 2003.

12. R. A. Finkel and J. L. Bentley. Quad trees: A data structure for retrieval on composite keys. *Acta Inf.*, 4:1–9, 1974.

13. K. Goh, B. Li, and E. Chang. Dyndex: A dynamic and nonmetric space indexer, 2002.

14. U. Guntzer, W.-T. Balke, and W. Kiesling. Towards efficient multi-feature queries in heterogeneous environments. *itcc*, 00:0622, 2001.

15. A. Guttman. R-trees: A dynamic index structure for spatial searching. In *SIGMOD'84, Proceedings of Annual Meeting, Boston, Massachusetts, June 18-21, 1984*, pages 47–57, 1984.

16. K. Kummamuru, R. Krishnapuram, and R. Agrawal. On learning asymmetric dissimilarity measures. In *ICDM*, pages 697–700, 2005.

17. N. Mamoulis, K. H. Cheng, M. L. Yiu, and D. W. Cheung. Efficient aggregation of ranked inputs. In *ICDE*, page 72, 2006.

18. T. Mandl. Learning similarity functions in information retrieval. In *EUFIT*, pages 771–775, 1998.

19. A. Marian, N. Bruno, and L. Gravano. Evaluating top- queries over web-accessible databases. *ACM Trans. Database Syst.*, 29(2):319–362, 2004.

20. J. Nievergelt, H. Hinterberger, and K. C. Sevcik. The grid file: An adaptable, symmetric multikey file structure. *ACM Trans. Database Syst.*, 9(1):38–71, 1984.

21. D. Padmanabhan and P. Deshpande. Efficient rknn retrieval with arbitrary non-metric similarity measures. *PVLDB*, 3(1):1243–1254, 2010.

22. D. Padmanabhan, P. M. Deshpande, D. Majumdar, and R. Krishnapuram. Efficient skyline retrieval with arbitrary similarity measures. In *EDBT*, pages 1052–1063, 2009.

23. H. Samet. *Foundations of Multidimensional and Metric Data Structures (The Morgan Kaufmann Series in Computer Graphics and Geometric Modeling)*. Morgan Kaufmann Publishers Inc., San Francisco, CA, USA, 2005.

24. T. Skopal. On fast non-metric similarity search by metric access methods. In *EDBT*, pages 718–736, 2006.

25. M. Theobald, G. Weikum, and R. Schenkel. Top-k query evaluation with probabilistic guarantees. In *VLDB*, pages 648–659, 2004.

26. J. K. Uhlmann. Satisfying general proximity/similarity queries with metric trees. *Information Processing Letters*, 40(4):175–179, 1991.

27. E. Vidal. New formulation and improvements of the nearest-neighbour approximating and eliminating search algorithm (aesa). *Pattern Recognition Letters*, 15(1):1–7, 1994.

28. D. Xin, J. Han, and K. C.-C. Chang. Progressive and selective merge: computing top-k with ad-hoc ranking functions. In *SIGMOD Conference*, pages 103–114, 2007.

29. P. N. Yianilos. Data structures and algorithms for nearest neighbor search in general metric spaces. In *Proceedings of the Fourth Annual ACM/SIGACT-SIAM Symposium on Discrete Algorithms, 25-27 January 1993, Austin, Texas.*, pages 311–321, 1993.

30. P. Zezula, G. Amato, V. Dohnal, and M. Batko. *Similarity Search - The Metric Space Approach*, volume 32 of *Advances in Database Systems*. Kluwer, 2006.

Chapter 7
The Road Ahead

As we have seen through this book, there has been a lot of recent work in enabling newer forms of similarity search by defining novel operators. Arguably, the search for novel similarity search operators are as much an active area of research as indexing and efficiency considerations in similarity search systems. We now outline the choices available in similarity search, under various heads:

- **Feature Modeling:** In many cases, it is desirable to bring multiple kinds of objects within the same similarity search system. A simple example is that of searching over profiles of service providers and consumers, as we have seen in bichromatic operators like *BRkNN* [10]. In certain cases, it is not so trivial to find a shared attribute space that incorporate the multiple kinds of objects. Consider the example of searching over restaurants and consumers, where the type of cuisine offered by the restaurant may be mapped to the same attribute as the cuisine preference of the consumer; while the former may be specified explicitly, the latter may need to be inferred implicitly by mining through recent restaurant visits by the consumer, or the social network profile of the consumer. In the case of a job-related search system, the job description, student profiles and job-seeker profiles all need to be mapped to the same schema.
- **Attribute-specific Similarity Estimation:** As we have seen in the similarity search framework, the similarity between two objects on each attribute is usually quantified using a numeric value. There are various choices for arriving at the numeric value depending on the nature of the attribute. If the attributes are string-valued, there are a variety of edit-distance adaptations to choose from. For values that are nodes in a network, choices include the number of intervening hops or random walk distance (e.g., [3]). Though these determine the construction of just the $s(q,x)$ vectors, and do not affect the downstream processes that form the core of the similarity search operator, different choices of similarity measures could yield search systems that yield significantly different results; this makes them an important decision choice.
- **Aggregation Functions:** This refers to the variability in doing the following operator-specific transformation:

© The Author(s) 2015
D.P and P.M. Deshpande, *Operators for Similarity Search,*
SpringerBriefs in Computer Science, DOI 10.1007/978-3-319-21257-9_7

$$s(q,x) \xrightarrow{\mathcal{O}} s_{\mathcal{O}}(q,x) \qquad\qquad (7.1)$$

The main choice is as to whether $s_{\mathcal{O}}(q,x)$ should be a single numeric value or a vector. If it is a vector, there is a further choice to determine the length of the vector; one possibility is to retain the $s(q,x)$ vector as is (e.g., skyline), or do some minor transformation to create a smaller vector (e.g., dynamic skyline [11]). Once the nature of the transformed representation has been decided, one needs to still decide on how to do the computation. Even for a single numeric value format for $s_{\mathcal{O}}(q,x)$, a plurality of simple choices such as *Min*, *Max* and *n-match* exist. In certain cases, we might want to use a more complex formulation that assigns higher weights to larger valued elements of the $s(q,x)$ vector (e.g., KNDN operator [8]). More choices exist in the case where the transformed representation is a vector.

- **Choice of Selection or Ranking:** Moving forward to the result estimation step, two high-level choices are as to whether to use a crisp binary result membership function (e.g., skyline, range query etc.) or to use a ranking of objects followed by a filter operation such as in the case of the weighted sum top-k operator.
- **Additional Features:** Having determined the choices of the basic operator in the above steps, one could enhance the operator by using one or more features such as *indirection*, *multi-query*, *subspaces*, *visibility*, *bichromaticity* and *diversity*. Many of these features can significantly alter the semantics of the base operator; for example, *indirection* provides a data object centric view rather than the query centric view provided by most basic operators.
- **Data Characteristics:** Another choice is on whether the semantics of the similarity search operator needs to be extended to work with uncertaintites or approximations in the data. For example, it is very common to not have the exact age value in a person database and instead have an age range. The exact disease may not be recorded for privacy reasons in a medical record database and instead the anonymized data may simply have a disease family. In other scenarios, the query may have a spatial spread in geographic data due to organic reasons such as the query being intrinsically a building that has a spatial extent [12].
- **Exact or Approximate:** For complex operators, clever indexing and search algorithms may still be inefficient; in such cases, exact algorithms for similarity search on an operator may not be feasible for use cases such as interactive querying. To handle those cases, one could come up with approximation algorithms that yield an approximation of the original result set much faster. These approximation algorithms may either be devised to provide theoretical guarantees on the level of approximation achieved, or may simply be empirically verified for approximation quality.

Differences in choices in any one of the above steps would yield search systems with varying semantics. The abundance of choices in each of the above phases have kept similarity search researchers very busy, and we might realistically expect that the activity in this space would continue in the forseeable future. We will now outline certain possible research directions in advancing similarity search systems. While some of these are very specific to new types of querying as has been the focus of this book, others address some other aspects of similarity search systems covering engineering/systems issues in building similarity search systems and new directions motivated by insights from disciplines outside computing.

7.1 Multi-modal Similarity Search

We had seen that the mind has an extra-ordinary capability of doing multi-modal similarity search across seemingly very disparate data sources; the Bouba-Kiki effect[1] discussed in the introduction illustrates that it is fairly trivial for the mind to identify seemingly sophisticated correlations between visual and lexical cues. Another related capability of the human mind that makes multi-modal similarity search simple is that of *intensity matching* [9]. This says that it is often perceived to be easy to match intensities and make judgements; for example, a question *Is X as tall as Y is intelligent?* is not perceived as very difficult for humans, whereas it would require sophisticated specialized statistical machinery to design a system to answer such a question. While creating a similarity search system that can do as well as the human mind may be a very far-fetched target, we believe that the following could be potential steps towards that large goal.

Modality-neutral Features: By observing neural activity patterns across subjects while exposed to the same kind of objects and by using the assumption that the same mental feature would generate similar responses across human subjects, one could identify a set of features that the mind uses, in identifying objects. These features may not necessarily map to individual attribute-values, but could map to value combinations across attributes. This exercise could be done on a per-domain basis (e.g., on just pictures), and then the identified features could be correlated to features in other domains (e.g., music). Such efforts could ultimately lead to cross-domain and modality-neutral features, that could be easily used in a cross-modal similarity search system.

Grounding to Modalities: Having identified some modality-neutral features, it would be interesting to develop a technology to ground an instance in that feature space to a real instance in a particular domain. When a domain is understood to the extent that an instance of the abstract feature space can be scored with respect to instances on that domain, it has been implicitly brought under the purview of the multi-modal similarity search system. In effect, the knowledge to ground instances in the abstract feature space to a particular domain gets rid of the need to do costly collection and mining of neural response data.

---

[1] http://en.wikipedia.org/wiki/Bouba/kiki_effect - Accessed March 22nd, 2015

7.2 Interoperability between Specialized Indexes

There are specialized indexes for various data types; these include inverted indexes that are popular for indexing text documents, and tree-structured indexes for trajectory data (e.g.,TrajTree [13]). These are designed for similarity search on respective data-types and are optimized for querying where queries are also of the same data type. However, text documents and user trajectories could appear as specific attributes of people entities in similarity search systems that search over multiple types of people entities such as candidate profiles (with publications being text documents) and user profiles for surveilance (where recent movements are captured as trajectories). Thus, instead of just finding the top-few similar documents or trajectories, we would like to find top-k similar entities where similarities on each of these attributes provide input to the pairwise object similarity computation. Efficient similarity search in such scenarios would require interleaving of the search operations across indexes in such a way that the final top-k similar entities be retrieved efficiently. Similar situation arises when distances on some attributes are Euclidean, whereas others are metric or non-metric. Different indexes need to be used on different attributes depending on the properties of their distance function. Inventing methods to do efficient and correct interleaving of search operations across different indexes would require an in-depth understanding of the indexes and their operation.

7.3 Explaining Similarity Search Results

Most common similarity search systems do not offer an explanation for inclusion or exclusion of objects in the result set. However, explanations are becoming increasingly critical in recommendation systems (e.g., [14]). Infact, popular email systems have started to provide explanations for certain classifications like *important messages*[2]. For document data, word-level rules have been explored to provide a semantic interpretation for a cluster [1]. Answering why certain objects have not been included in the results, i.e., *why not* questions, has been a theme that has been attracting recent interest in the context of top-k queries [6], reverse top-k queries [4] and reverse skyline queries [7]. In general, explaining similarity search results would be an interesting frontier for research in similarity search systems.

7.4 Leveraging Newer Platforms for Similarity Search

With the dataset sizes increasing to levels that reach limits of a single system, we would need to have similarity search algorithms running on platforms such as

---

[2] http://email.about.com/od/gmailtips/qt/How-To-Find-Out-Why-Gmail-Categorized-A-Message-As-Important.htm - Accessed March 22nd, 2015

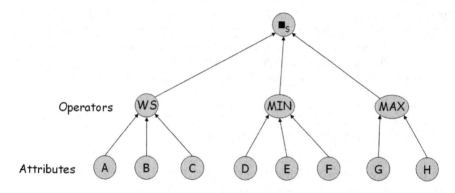

Fig. 7.1: Operator Tree.

Hadoop/MapReduce soon. MapReduce platforms have been explored for join processing quite extensively in the last few years (e.g., [5]). There have been efforts in doing domain-specific similarity search, such as similarity search on image data[3], on the MapReduce framework. Adapting similarity search algorithms to work on such new platforms that enable easy scaling are a useful direction for impactful work towards building practical large-scale similarity search systems.

7.5 Complex Operator Combinations

In the case of similarity search on complex object types that have a wide variety of attributes in their schema, it might be necessary to have complex combinations of operators each operating on subsets of attributes, to address similarity search scenarios. As an example, let us consider the two basic operators, *weighted sum* and *skyline*; the weighted sum operator allows for dissimilarity on one attribute to be compensated by high similarity on other. On the other hand, the skyline operator considers each attribute as being independent of the other. A complex object schema could potentially be divided into groups of attributes, of which some groups may have attributes that are *compensatory* whereas other groups contain mutually *independent* attributes; such groups may be handled by weighted sum and skyline operators respectively. The weighted sum or some other operator may be used to aggregate across results from such group-level operators. Figure 7.1 illustrates the case of an 8 attribute schema, that are divided into three sets; the first set of three attributes are aggregated using the weighted sum operator, while the second and third sets, of three and two attributes each, have the Min and Max operator handling them. The three outputs are then fed to a top-level skyline operator. Infact,

---

[3] http://www.slideshare.net/denshe/terabytescale-image-similarity-search-with-hadoop - Accessed March 25th, 2015

the *dynamic skyline* operator handles this kind of an operator tree for the special case where the tree is shallow, and where the skyline operator should necessarily be the top-level operator. Generalizing such a framework to general operator trees and designing efficient algorithms to process such operator trees would be technically challenging; [2] discusses some thoughts in that general technical direction.

7.6 Novel Similarity Search Operators

As has been emphasized earlier in this chapter, there is an abundance of design choices available towards building a similarity search system. Of prime consideration to us, being the focus of this book, are the choices available to design newer similarity search operators. It is often simple to define a new similarity search operator, by

- Choosing a basic operator and an optional filter condition, and
- Choosing additional features to use on the basic operator.

Many of the features we have considered in the earlier chapters are compatible with each other, allowing to define an operator by choosing multiple features. Uptake of operators, however, depend on other factors such as whether there are real-world scenarios targeted by the new operator that cannot be adequately addressed by existing operators, and the availability of efficient algorithms to process the operator in question. Due to the technical complexity in ensuring efficiency, algorithms and indexing for efficient computation has been the target of most research in similarity search. We now outline a few novel operators along with motivating scenarios, to illustrate that there is much space for newer similarity operators.

7.6.1 N-Match-BB: Any N conditions

We will now define the semantics of the result set of this new operator in the familiar declarative fashion:

$$N\text{-}Match\text{-}BB(q, \mathscr{X}) = \{x | \exists A \subseteq \mathscr{A}, |A| = N \wedge (\forall a \in A, d(q,x)[a] \leq \tau[a])\} \quad (7.2)$$

where $d(q,x)[a]$ is an overloaded notation indicating the distance between q and x on the attribute a. τ is a vector that specifies the threshold of distance that is required on each attribute in $\mathscr{A}$. Informally, this chooses all objects in $\mathscr{A}$ that satisfies the maximum distance threshold on at least N attributes in $\mathscr{A}$.

Example: Figure 7.2 illustrates the simple example of a 2-d attribute space with a query. The grey boxes show the areas that satisfy the distance threshold on each of the attributes. In this simple case, a *1-Match-BB* operator would return all objects

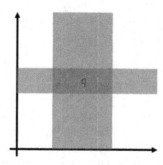

Fig. 7.2: Illustration of the N-Match-BB Operator.

that fall on either grey box. However, the result set becomes more interesting when it come to three attributes or more.

Motivation: This operator is motivated by eligbility determination scenarios in contexts such as job search or applying for a governmental benefits scheme. In such cases, there are typically multiple conditions of which at least a pre-specified number of conditions need to be satisfied by a person to be considered eligible. The conditions on each attribute can be formulated as a range condition, wherein the eligibility question becomes one of membership in the *N-Match-BB* result set.

7.6.2 Multi-Query BRkNN: Multi-Query Extension of BRkNN

This operator adds the multi-query feature on the *BRkNN* operator that we have discussed earlier. The result set is defined as the following:

$$MQ\text{-}BRkNN(Q, \mathscr{X}) = \{x | x \in \mathscr{X}_{type \neq Q.type} \wedge (Q \cap top\text{-}k(x, \mathscr{X}_{type=q.type} \cup \{Q\})) \neq \phi\}$$
$$(7.3)$$

where $Q.type$ is the type of the query objects; all the query object need to be of the same type. The result set of the *MQ-BRkNN* operator is the set of objects that have at least one of the query objects in their respective top-k result sets. The $(Q \cap top\text{-}k(x, \mathscr{X}_{type=q.type} \cup \{Q\})) \neq \phi$ predicate becomes true only in those cases when the intersection of the top-k result set with the query set is not null, consequently implementing the semantics outlined above.

Motivating Example: Consider the case of a cafe chain that is considering a set of locations to start outlets in a city. This cafe chain would like to estimate the number of potential customers across all locations, when just the geographical proximity is considered. Assuming that each consumer would typically visit only the top-k cafes that are closest to her, the *MQ-BRkNN* query estimates the number of consumers who would find one of the outlets (i.e., if outlets were opened in all the k candidate locations) among their closest k sets of cafes.

7.6.3 Summary

As discussed above, it may be seen that there are similarity operators that are useful for intuitive real-world scenarios that are yet to be proposed and studied. There is a lot of scope for being creative in defining new features and new combinations that would lead to new operators not discussed in this book. Devising new operators and efficient algorithms to implement such operators in practical systems is a potential research direction that holds much promise to enhance current similarity search systems.

7.7 Summary

In this chapter, we first started by looking at the various choices available for the designer of a similarity search system, categorized according to the stage in the similarity search pipeline that the choice is applied. We then discussed various potential research directions in advancing similarity search systems. We started by outlining some preliminary ideas on enabling multi-modal similarity search, and later discussed algorithmic challenges in coping with multiple kinds of indexes within a similarity search system. Explaining similarity search results, a relatively new consideration in similarity search, was also examined as a potential research direction. We observed that the inevitable next step in scaling similarity search systems would be to develop methods for similarity search on the Hadoop/MapReduce framework that is becoming popular of late. We also considered similarity search where attributes are combined through an operator tree since a single similarity search operator may not be sufficient to address novel scenarios over complex data. Turning our attention to research in developing novel similarity search operators, we illustrated that the conceptualization of operators as basic operators with features, outlined in an earlier chapter, makes it easy to design novel operators. In making it evident by means of examples, two novel operators were outlined along with motivating scenarios where they could be used. We hope that this chapter was useful in providing insights to identify and prioritize directions for research in advancing the frontier in similarity search, for interested researchers.

References

1. V. Balachandran, P. Deepak, and D. Khemani. Interpretable and reconfigurable clustering of document datasets by deriving word-based rules. *Knowledge and information systems*, 32(3):475–503, 2012.
2. P. Deshpande, R. Krishnapuram, D. Majumdar, and D. Padmanabhan. Retrieval of relevant objects in a similarity, Dec. 30 2010. US Patent App. 12/491,485.
3. T. Emrich, M. Franzke, N. Mamoulis, M. Renz, and A. Züfle. Geo-social skyline queries. In *Database Systems for Advanced Applications*, pages 77–91. Springer, 2014.
4. Y. Gao, Q. Liu, G. Chen, B. Zheng, and L. Zhou. Answering why-not questions on reverse top-k queries. *Proceedings of the VLDB*, 501:1, 2015.
5. H. Gupta, B. Chawda, S. Negi, T. A. Faruquie, L. V. Subramaniam, and M. Mohania. Processing multi-way spatial joins on map-reduce. In *Proceedings of the 16th International Conference on Extending Database Technology*, pages 113–124. ACM, 2013.
6. Z. He and E. Lo. Answering why-not questions on top-k queries. *Knowledge and Data Engineering, IEEE Transactions on*, 26(6):1300–1315, 2014.
7. M. S. Islam, R. Zhou, and C. Liu. On answering why-not questions in reverse skyline queries. In *Data Engineering (ICDE), 2013 IEEE 29th International Conference on*, pages 973–984. IEEE, 2013.
8. A. Jain, P. Sarda, and J. R. Haritsa. Providing diversity in k-nearest neighbor query results. In *Advances in Knowledge Discovery and Data Mining*, pages 404–413. Springer, 2004.
9. D. Kahneman. *Thinking, fast and slow*. Macmillan, 2011.
10. F. Korn and S. Muthukrishnan. Influence sets based on reverse nearest neighbor queries. In *ACM SIGMOD Record*, volume 29, pages 201–212. ACM, 2000.
11. D. Papadias, Y. Tao, G. Fu, and B. Seeger. Progressive skyline computation in database systems. *ACM Transactions on Database Systems (TODS)*, 30(1):41–82, 2005.
12. R. Pereira, A. Agshikar, G. Agarwal, and P. Keni. Range reverse nearest neighbor queries. In *KICSS*, 2013.
13. S. Ranu, D. P, A. Telang, S. Raghavan, and P. Deshpande. Indexing and matching trajectories under inconsistent sampling rates. In *31st International Conference on Data Engineering*, 2015.
14. N. Tintarev. Explanations of recommendations. In *Proceedings of the 2007 ACM conference on Recommender systems*, pages 203–206. ACM, 2007.

Index

Printed in the United States
By Bookmasters